Democratic Management of an Ecosystem under Threat

Democratic Management of an Ecosystem under Threat

The People's Reefs

Kelly Dunning, Amanda Alva, Sabine Bailey,
Daniel Morris, Kasen Wally, and Ryan Williamson

ANTHEM PRESS

Anthem Press
An imprint of Wimbledon Publishing Company
www.anthempress.com

This edition first published in UK and USA 2023
by ANTHEM PRESS
75–76 Blackfriars Road, London SE1 8HA, UK
or PO Box 9779, London SW19 7ZG, UK
and
244 Madison Ave #116, New York, NY 10016, USA

British Library Cataloguing-in-Publication Data
A catalogue record for this book is available from the British Library.

Library of Congress Control Number: 2023932898
A catalog record for this book has been requested.

ISBN-13: 978-1-83998-671-0 (Hbk)
ISBN-10: 1-83998-671-9 (Hbk)

Cover Credit: Sabine Bailey. "Florida Lauderdale by the Sea" and "Florida Keys Reef" images, The Conservation Governance Lab, College of Forestry, Wildlife, and Environment, Auburn University.

This title is also available as an e-book.

CONTENTS

Acknowledgments: The authors thank their friends and family for support in putting together this book. Thank you to Dr. Larry Susskind for the early guidance that ultimately became this book.

Funding acknowledgments: Funds for this research have been provided by a Hatch and Multistate Project through the Alabama Agricultural Experiment Station (AAES) grant number ALA0- 31-1-19098; Funds for this research have been provided by the Early Career Faculty Innovator Program at the National Center for Atmospheric Research, a program sponsored by the National Science Foundation. Any opinions, findings and conclusions or recommendations expressed in this material do not necessarily reflect the views of the National Science Foundation. Cooperative Agreement Number: 1755088; Funds for this research have been provided by the Early-Career Research Fellowship from the Gulf Research Program of the National Academies of Sciences, Engineering, and Medicine. The content is solely the responsibility of the authors and does not necessarily represent the official views of the Gulf Research Program of the National Academies of Sciences, Engineering, and Medicine. Grant number 250147-145001-2000; Funds for this research have been provided by the National Oceanic and Atmospheric Administration (NOAA) Climate Programs Office Grant number 202281-145001-2000.

Part 1

THE CLIMATE CHANGE CHALLENGE TO CORAL REEFS THAT WILL REQUIRE CONSERVATION THEORY AND PRACTICE TO EVOLVE

Chapter 1

INTRODUCTION

By Kasen Wally and Kelly Dunning

In 2019 I was working for a small non-profit on the northern coast of Haiti, near the port city of Cap-Haïtien. Historically, Cap-Haïtien was nicknamed *The Paris of the Antilles* for its great wealth and beauty as a former French Colony (see figure 1.1) (Liss & Knight, 1991). Today, Cap-Haïtien remains quite beautiful, retaining much of its colonial-era architecture, which is now steeped in vibrant Haitian culture and nestled between lush mountains and a sprawling turquoise bay. My work there involved coral reef restoration, or replanting coral to help bring the reef back to its former glory, and thus the brilliant blue waters are where I spent most of my time. Despite Haiti's rugged beauty, environmental degradation is abundant, and its coral reefs are heavily impacted. Much of the loss of coral reefs is due to there being a lack of ways to manage waste, pollution, overfishing, coastal development and a lack of cohesive environmental *governance*, which we define as political management combining government, non-profits, public participation and private-sector interests (Creary et al., 2008).[1] Governance and government are not the same things, and this distinction is necessary for our book's central arguments to unfold. Government intervention includes actions like a fish and wildlife agency instituting catch-limit regulations on a fishery, whereas governance includes the activities of other actors such as communities, businesses and non-governmental organizations or NGOs (Lemos & Agrawal, 2006).

On an otherwise tranquil morning dive, I began to witness these impacts firsthand. A large shadow passed slowly over me, which turned out to not be a floating patch of *Sargassum*, a genus of seaweed commonly found in the Caribbean. Instead, it was an enormous floating mass of single-use plastics, such as the small yellow plastic vinegar bottles typical of Haiti (see figure 1.2).

1 This information was collected on the western coast of Haiti and thus may differ from that of Haiti's northern coral reefs.

Figure 1.1 View from coral survey field site on Amiga Island in Baie de I'Acul, Haiti. Image courtesy of Kasen Wally.

This marine debris had washed down from the mountains during a storm the night before, and was symbolic of the immense challenges facing Haitian coral reefs. Beachgoers as far as Corpus Christi, Texas, in the United States commonly find these same yellow vinegar bottles washed up on Gulf of Mexico beaches nearly 3,000 kilometers away, with missing diamond shapes, from bites taken from the bottles by sea turtles. The interconnectedness between coastal communities can often be traced in such a way.

In the weeks that followed, I witnessed some of the most severe coral bleaching I have ever seen due to the unusually warm ocean temperatures at the time. Frustratingly, many of the corals we were studying and transplanting as part of restoration work perished, and I was left wondering, "How can we truly address these problems and make a difference when changes are happening so rapidly?" Despite my firsthand experience, the current status of Haiti's coral reefs, especially in the north, officially remains unknown (Creary et al., 2008). However, we do know that Haiti's coral reefs are not alone; worldwide, coral reefs are under severe pressure from emerging threats, most of which are anthropogenic, or caused by the activities of people. But, there is hope for coral reefs, and while many of the proposed solutions are not necessarily simple, they are rapidly evolving. Part 1 of our book introduces the ongoing and emerging pressures facing coral reefs and further explores the theory and practice of what we call *participatory management*, or coral reef

Figure 1.2 Vinegar bottles from Haiti collected on Corpus Christi coasts by Mission-Aransas Reserve Manager Jace Tunnell. Image courtesy of Jace Tunnell.

management that involves the people who depend most intensely on the reefs. It is this form of management which we argue will help shepherd reef management through the next century.

Changes in recent decades to the way that we protect coral reefs have been in response to increasing threats on a global scale (Chaffin et al., 2014; Hughes, 1994; Morrison et al., 2020). While climate change and ocean acidification are the most significant threats facing modern coral reefs, other human-made, or *anthropogenic* stressors such as pollution, overfishing and physical damage (e.g., ship groundings, oil drilling, anchor damage and trawling) threaten the health and productivity of coral reef ecosystems, and serve as the original causes for people to set up political arrangements aimed at coral reef conservation (Frys et al., 2020; Hughes et al., 2017b).

Realizing that reefs were on the brink, conservationists and decision-makers set up some of the first systems of rules, organizations and norms for coral reef management to respond to the stressors like overfishing and vessel groundings. We call these systems of rules and norms *institutions* for managing ecosystems, and they are particularly helpful when ecosystems are facing threats. Institutions are one of the most important topics of this book, and are formally defined as the systems structured by accepted social norms in a particular place, as well as rules governing access to a natural resource which can be reorganized via humans making those decisions (Crawford & Ostrom, 1995). Institutions can include organizations, government agencies and places of worship; all with an emphasis on shared rules and regulations.

Institutions are primarily created to help solve social issues and dilemmas (Imperial & Yandle, 2005).

Most of what we know about effective institutions for coral reef management comes from the immense legacy of Dr. Elinor Ostrom, whose theories of institutions grounded in rigorous field work in real communities examined how people work together to manage common pool resources. Common pool resources are those that are hard to exclude users from (e.g., clean air and water) and where conventional wisdom would lead one to believe that people instinctually would exploit such resources until they had vanished. Ostrom's work asserted that this conventional wisdom was wrong, citing numerous examples of communities, many of which were low-income communities, coming together to make decisions to manage their natural resources and ensure future generations had access to scarce natural resources. She specifically described the core design principles for effective management of natural resources, conceptualizing these principles as institutions for sound resource management (Ostrom, 1990). Ostrom's Institutional Analysis and Development framework places institutions in a wider context that includes the biophysical ecosystem (e.g., coral reefs), attributes of the local and global community (e.g., the many stressors reefs face from industry across sectors), interactions between individuals making decisions to conserve reefs within organizations and the ecological outcomes that result (Ostrom, 2011). What Elinor Ostrom did for people thinking about how we can best protect coral reefs was to connect questions of how people organize themselves to make decisions to the ecological outcomes (desired or undesired) that result. The majority of articles on institutions for managing coral reefs draw on, test and build on Ostrom's frameworks by analyzing the relationships between social, economic and political data (using institutional frameworks) and ecological data (see Cinner et al., 2012; Cinner & Huchery, 2014, *ibid.*).

Studies on effective management institutions for coral reefs are numerous, and briefly, have findings that suggest that community participation in management and flexible institutions that can be adapted to local people and culture are related to healthy reef ecosystems (Beger et al., 2004; Cinner et al., 2016; Cinneret al., 2009; Cinner et al., 2016); that desirable ecological variables such as fish biomass are related to management institutions that invite every day people to play a role in decision-making and institutions that leverage local culture and traditions in management (Cinner et al., 2016; Ferse et al., 2014). Institutions alone may not be the only factors related to reef ecology, external factors like nearby markets, local wealth and population size may negatively impact conservation efforts (Cinner et al., 2016; Ferse et al., 2014). Research also shows that large conservation protected areas and institutions that invite local people to act as decision-makers can be combined for effective conservation (McClanahan et al., 2006); that decision-making power shared with local communities, called *co-management institutions*, can meet social and ecological goals while still succumbing to pitfalls such as primarily benefiting the wealthy (Cinner et al., 2005, 2012; Cinner et al., 2005); that integrating customary management practices strengthens conservation efforts (Cinner & Aswani, 2007); that historical and social variables are important when analyzing implementation of reef management institutions (Cinner et al., 2009; Wells et al., 2016); that conservation of protected area boundaries based on social relations and local cultural identity are important to consider in reef protection (Hicks & Cinner, 2014); that community educational programs are related to social and ecological success (Pietri et al., 2009); and that government support enhances the effectiveness of conservation efforts (Léopold et al., 2019).

With hundreds of articles written on institutions for managing coral reefs, there is near universal agreement on one finding: institutions for coral reef management must be flexible and able to adapt in the face of change (Chaffin et al., 2014). In the past decade, research on the need for flexible institutions has increased due to an emerging understanding that climate change impacts are increasing, making the likelihood of returning reefs to configurations of the past no longer possible in an era known as the anthropocene (Cinner et al., 2012; Hughes et al., 2017).

For coral reefs, marine protected areas (MPAs) are the most important institutions. MPAs are conservation areas under the water with rules governing activity impacting the ecosystem, such as limits or bans on fishing, and strict standards for anchoring vessels. MPAs became a standard coral reef management tool in the United States as early as 1970 with the creation of the federal agency tasked with protecting American marine resources including reefs, the National Atmospheric and Oceanic Administration (NOAA). Even earlier than this federal effort to protect coral reefs, the state of Florida established John Pennekamp Coral Reef State Park in 1963, an MPA which, combined with the Florida Keys National marine Sanctuary[2] was approved by former president Gerald Ford in 1975. As an MPA, Pennekamp Park limits fishing, touching reef species and harvesting key marine species on reef ecosystems across 3800 square miles. These MPAs in South Florida embody traditional forms of coral reef management: governmental decision-makers

2 Designated as the Key Largo National Marine Sanctuary in 1975, and designated by the United States Congress as the Florida Keys National Marine Sanctuary in 1990.

prioritizing a spatial extent of coral reef ecosystems in need of protection, creating rules for exploitation involving fishing and other uses, enforcing rules, and to varying extents, engaging the interested groups of people, or stakeholders, who are impacted by these decisions. This type of management is known as top-down management, but it is not the only way to protect coral reefs.

Beginning in the 1990s studies began to describe the beginnings of conservation governance changes in MPAs focused on power sharing with local communities in Southeast Asian settings (Pomeroy, 1995) and the institutional changes needed to empower communities to manage local stressors such as destructive fishing (Christie et al., 1994). In many ways, the beginnings of people-centered conservation began in the Global South, where communities like villages were granted decision-making power over coral reef conservation, legitimized and shared with the central government.

These early studies make an argument for decentralized coral reef management with resources users engaged in what has been referred to as *cooperative management*, *collaborative management*, or *community-based management*, where people that use the reef, like fishermen or dive boat operators, are also the managers that govern its protection. Community-based management is not a prescription, but a localized process that starts from the ground up and involves networks and linkages across various levels of organization, or scales, with the collaboration of local and national governments, NGOs and donors (Berkes, 2010).

When you combine elements of early MPAs created by the government and power sharing with local communities, formal power sharing is put in place through legal recognition and government support. This type of collaborative management is termed *co-management*, with several added benefits such as conflict reduction between users, learning about best practices in conservation, and even reduced food insecurity (Pomeroy & Rivera-Guieb, 2009). Despite the benefits outlined in case studies and in theory, and touted by international funders and governments, some investigators point out that in many cases, devolution of the 1990s failed due to the very dynamism that makes collaborative management so appealing in the first place (Berkes, 2010; Blaikie, 2006; Evans et al., 2011). Cinner et al. expand on the weaknesses in their in-depth examination of 42 co-management arrangements in five countries that found that while co-management can meet many social and ecological goals, it does not impact ecosystem conditions as much as stronger factors like market access and user dependence on resources (2012). In other words, if there is an economic incentive to destroy reefs, it does not matter what governance arrangement (top-down, co-managed, etc.) is in place as these factors will supersede management institutions.

Beginning in recent decades, anthropogenic stressors have become so common and so detrimental that entire coral reef ecosystems have entered a novel state of existence, where groups of species, their interactions and their functions or roles in nature have been permanently altered (Graham et al., 2014, Hughes et al., 2017a). In combination with high levels of uncertainty over how severe the changes to the climate will be, and what the accompanying impacts will look like, traditional management is failing to address these novel conditions. For example, the US state of Florida has lost 90 percent of its coral under these traditional management institutions (Voosen, 2019). Instead, and in the face of compounding stressors, a new requirement for collaboration across diverse interests, regions and institutions has emerged (Graham et al., 2014; Schultz et al., 2015). Yet to date, there has been little written on how new and emerging institutions that center participation of stakeholders have the potential to respond to climate change and novel ecosystems.

The need for this book first became clear to me when diving a stretch of one of the most popular reefs in the Florida Keys National Marine Sanctuary. Looking around me, I noticed all of the hard coral was dead, with the only remaining hard coral being that which had been replanted by the patchwork of Florida-based companies and non-profit organizations that make reef restoration the focus of their work. Swimming up to a cluster of replanted staghorn coral (*Acropora cervicornis*), words that typically can be found only in academic papers such as "novel ecosystem" began to take on a more concrete meaning right before my eyes. To what degree must the way that we steward coral reefs change given how much damage these systems have already absorbed. Looking down at the replanted corals, labeled with colorful plastic tags, this question began to ring much more urgent to me than the core thesis of this book.

It is our hope that the types of highly adaptive, participatory management strategies that we study in this book may help guide conservation practitioners and policymakers toward desirable large-scale environmental outcomes (Folke et al., 2005; Berkes et al., 2008; Lockwood et al., 2012). Given the rapidly changing nature of *socio-ecological systems* (defined as the interactions between people and the environment which they depend upon) in the context of global environmental change, there is a clear need for transformative governance that builds ecological resilience while protecting the interests of coastal communities (Morrison et al., 2020; Pendleton et al., 2016; Rogers et al., 2015). Our book aims to describe, in a clear and accessible way, five case studies of this transformative, adaptive governance of coral reefs. Our book presents these case studies showing stakeholder-driven policy-making that includes the people who depend on the reef. Our case studies examine the national level through bipartisan legislation in the United States, at the sub-national level through participatory management in Florida and at the local level through grassroots movements in Florida and the Cayman Islands. Its contribution is aimed at policy-makers looking to replicate successes and learn from mistakes from other localities responding to anthropogenic and environmental change with every day people invited to participate with a seat at the table.

Coral Reef Ecology

Coral reefs are irreplaceable, rare and vulnerable ecosystems. They are described as "global life-support systems," sustaining at least 500 million people, with 25 percent of all marine species worldwide present on reefs (Moberg & Folke, 1999; Henkel, 2010). Although tropical coral reefs cover less than 0.5 percent of the ocean floor, they are among the most diverse and productive[3] ecosystems on the planet (Hoegh-Guldberg et al., 2017; Moberg & Folke, 1999). Coral reefs need specific conditions to thrive, including ideal water temperatures; appropriate salinity; exposure to sunlight; and clean, clear and low-nutrient waters (Hoegh-Guldberg et al., 2017). As such, these reefs are limited in their geographic location. They are generally found between the Tropic of Cancer (~30° north of the equator) and Tropic of Capricorn (~30° south of the equator), in shallow waters from 0 to 100 meters from the coast, and reside in the Pacific Ocean, Atlantic Ocean, Indian Ocean, Caribbean Sea, Red Sea and Persian Gulf (Hoegh-Guldberg et al., 2017). Furthermore, any drastic and long-lasting changes to some of these conditions can lead to widespread coral mortality, highlighting their vulnerability. Despite this, what we consider as modern coral reefs have been around for over 200 million years, first emerging in the late Triassic Period (Stanley, 1981).

Coral reefs are often characterized by their *scleractinian species*, otherwise known as "reef-building," "stony" or "hard" corals. As their colloquial descriptions imply, these species play a critical role in the formation of coral reefs (Stanley, 1981). Coral reefs are composed of hundreds of different reef-building species, and each coral colony is composed of thousands of genetically identical animals called polyps. Stony corals grow through a process

3 The productivity of an ecosystem refers to the rate of formation of biomass, which can occur from primary production (e.g., where plants are producing organic material from inorganic material through photosynthesis for example) as well as secondary production where biomass is generated by consumer organisms in a system.

of accretion, producing layered calcium carbonate skeletons that protect the delicate coral polyps, secure them to the substrate, and as a result, build and add structural complexity to the reef (Roff, 2019). Coral reefs create the largest biological structures on Earth. For instance, the Great Barrier Reef in Australia can be seen from space. Besides their vibrant colors and almost otherworldly appearances, it is these contradictions that make reefs so fascinating, how an organism as tiny as a coral polyp can, over centuries, build a geological formation visible from space.

The social and economic importance of coral reefs cannot be overstated. As life-support systems, coral reefs provide a litany of benefits to human well-being, benefits known as ecosystem services (Woodhead et al., 2019). Among the many ecosystem services coral reefs provide, they generate robust tourism industries, support vast commercial and recreational fisheries, hold deep cultural significance and protect coastlines during severe storm events. For example, a recent evaluation concluded that in the United States alone, coral reefs safeguard more than 18,000 people from coastal flooding every year (Storlazzi et al., 2019). Despite the beneficial relationships between coral reefs and coastal communities, the true socioeconomic impacts that the loss of these ecosystems would have are difficult to predict (Hoegh-Guldberg et al., 2019). For example, estimates of the global value of the provided goods and services (e.g. fisheries, coastal protection, etc.) range from $29.8 billion to $2.7 trillion annually (Conservation International, 2008; Hoegh-Guldberg, 2017; Scouter et al., 2021a;).

Chapter 2

CLIMATE CHANGE, CORAL BLEACHING AND OTHER THREATS

By Kasen Wally and Kelly Dunning

In recent decades, climate change has taken center stage as the single greatest threat facing modern coral reefs (Hughes et al., 2017a, 2017b). Atmospheric greenhouse gas concentrations have been increasing since the Industrial Revolution. The concentration of one commonly recognized contributor, carbon dioxide (CO_2), has risen at 250 times its natural rate following the last Ice Age (NASA, 2020). Currently, the measurement of atmospheric CO_2 stands at 415 ppm, the highest concentration in 650,000 years (NASA, 2020). As a result, the Earth's atmosphere and oceans are warming at an unprecedented rate (Hopkin, 2005). Over 90 percent of the warming that the planet has experienced in the last 50 years has occurred in the oceans themselves. From 1971 to 2010, 63 percent of the total increase in stored heat on our planet was captured by the oceans' upper, sun-lit layers. Therefore, the oceans' usefulness as a reservoir for storing thermal energy (heat) and CO_2 may spell disaster for coral reefs. Because coral reef ecosystems are among the most climate-sensitive on Earth, requiring narrow temperature ranges to survive, they are extremely vulnerable to unprecedented warming events.

One catastrophic effect of climate change, a phenomenon known as *coral bleaching*, has become a well-documented and highly publicized process (Dispensa et al., 2003; Hughes et al., 2017a, 2018a, 2018b, Manzello, 2015). Because corals are sessile organisms (i.e., they are immobile), they rely on two sources of energy. First, they capture microscopic organisms in the water column using small tentacles armed with stinging cells called *nematocysts*. Second, corals depend on a symbiotic relationship with several specific groups of *dinoflagellate microalgae* (i.e., single-celled algae) called *zooxanthellae*—these microorganisms produce oxygen and energy for the corals via photosynthesis, exactly like plants on dry land, in exchange for shelter and shared nutrients.

Zooxanthellae also give corals their often-vibrant coloration ranging from orange to yellow to royal blue. Because physically capturing microorganisms as food only provides corals with about 10 percent of their energy, the zooxanthellae are essential for their survival (Forsman, 2005; Weis, 2008). However, when coral organisms are exposed to adverse environmental conditions for prolonged periods of time (e.g., warming ocean temperatures), corals exhibit a natural stress response known as "bleaching." During this process, coral colonies expel the symbiotic zooxanthellae, revealing their underlying white calcium carbonate skeletons and losing these vital energy sources. If the environmental stressors like heat are not alleviated, coral bleaching events often result in high rates of coral death known as mortality events (Wagner et al., 2010). Unfortunately, due to global warming's rapid progression, these events are no longer isolated or rare (Hughes et al., 2003; Manzello et al., 2007). Nearly everyone who spends time under water enjoying coral reefs has witnessed a bleaching event, where corals appear stark white, beautiful if not haunting due to the underlying cause. While such events used to be regarded as rare, nowadays they are commonplace, another example of a novel ecosystem.

The extent and severity of coral bleaching are the result of above-average sea-surface temperatures and long-lasting temperature fluctuations that endure longer than what is typically found due to natural variability. While many corals are resilient to brief heat waves, sustained elevations in sea surface temperatures can cause high mortality in coral reefs. It has been shown that prolonged exposure to mean temperatures as little as ~1°C (1.8°F) above the regional summer maximum can lead to coral bleaching. As a result, coral colonies often perish before conditions stabilize (Hoegh-Guldberg, 1999; Hughes et al., 2003; Magel et al., 2019). In 2013, the Intergovernmental Panel on Climate Change (IPCC) predicted a 1–4°C (1.8–7.2°F) increase in mean global ocean temperatures by 2100—a bleak prognosis for coral reefs.

While climate change has been identified as the primary threat to coral reefs, its impacts are not limited to coral bleaching, nor are the threats to coral reefs limited to climate change. Another adverse impact that is closely related to climate change is ocean acidification which limits coral reef growth and damages other marine organisms. Much like the oceans' ability to store heat, it is also a sink for greenhouse gases, most notably CO_2 emissions. Between the industrial revolution and the mid-1990s, it has been estimated that the world's oceans absorbed 30 percent of all anthropogenic CO_2 and will continue to do so at similar rates (Gruber et al., 2019). To some, this might sound like a great thing, what could be better than an enormous CO_2 reservoir covering 71 percent of the Earth's surface? While the oceans do keep that CO_2

out of the atmosphere, the problem arises via a series of chemical reactions that occur when CO_2 is dissolved in seawater, causing the water to become more acidic (i.e., the pH of the water is reduced). When this happens, calcium carbonate minerals which are normally supersaturated in the seawater, are removed (i.e., become undersaturated) and thus cannot be utilized by organisms like corals and shellfish to produce their calcium carbonate skeletons and shells. For corals, this has been shown to damage reproductive success and may lead to a loss of coral cover (Albright, et al., 2010; National Oceanic and Atmospheric Administration, n.d.).

In other words, reef-building corals may ultimately be lost to ocean acidification faster than they can be replaced. Additionally, NOAA estimates that by the end of the twenty-first century, the oceans may become more acidic than at any time in the last 20 million years (National Oceanic and Atmospheric Administration, n.d.). Once again, much like atmospheric CO_2, this is an alarming threat, pointing to a global shift toward conditions never seen by any human on planet Earth.

In addition to thermal stress and ocean acidification, a recent study found that water pollution, fishing pressure, rising coastal populations and industrial development were the top-ranked cumulative pressures on the world's coral reefs (Andrello et al., 2021). Though the extent and severity of these stressors differ at regional scales, they build on previous results which have also identified destructive fishing practices (such as the use of dynamite and cyanide to remove key species from the reef), coral disease and invasive species as significant causes for concern (Alvarez-Filip, 2019; Burke et al., 2012; Hoegh-Guldberg, 2019).

Research also indicates that these stressors are typically synergistic (i.e., the impacts of coral disease are worsened by climate change and further compounded by overfishing), happening concurrently and exacerbating coral reef decline (Dustan, 2003). For example, along the Florida Keys Reef Tract, the third largest barrier reef in the world, the once pristine coral cover decreased for a variety of reasons: (1) the Florida Reef Tract's proximity to population centers like the city of Miami; (2) the frequent hurricane encounters on the reef; (3) fishing pressure from recreational and commercial fishery interests; (4) coastal development resulting in profound disturbances to water quality; (5) altered hydrology and nutrient flows from the Florida Everglades; and (6) climate change (Jackson et al., 2014; Precht & Miller, 2007). A world-renowned diving industry in the Florida Keys also places pressure on the reef, with visitors accidentally touching and disturbing the reef with every visit.

Chapter 3

STATUS OF CORAL REEFS

By Sabine Bailey and Kelly Dunning

These threats described in Chapter 2 have led to the tragic loss of nearly 50 percent of reef-building corals worldwide in the past 30 years, and their status continues to decline (Bruno & Selig, 2007; Côté et al., 2005; De'ath et al., 2012, Eddy et al., 2021; Hoegh-Guldberg et al., 2019). We are currently in a coral reef crisis with one-third of our reefs threatened with extinction (IUCN, 2021). According to the latest Global Coral Reef Monitoring Network report (2021),[1] from 1978 to 1997, the world's stony coral populations remained relatively stable, accounting for approximately 32 percent of reef cover, meaning that when you look at the seafloor when swimming along a reef, you see about one-third of the bottom covered in hard coral species. However, in 1998, a mass coral bleaching event resulted in the loss of 8 percent of that cover, though the reefs were able to rebound back to ~33 percent by 2009. Since 2009 however, there has been a steady decline in stony coral cover globally and a shift from coral-dominated reefs to algal-dominated ones, with various species of algae supplanting the space on the seafloor where the stony corals used to be (Scouter et al., 2021a). Algae and coral fight for the same habitat as both need similar conditions, especially exposure to sunlight (Swierts & Vermeij, 2016). From 2009 to 2018, we lost 13.5 percent (4,579 square miles) of stony coral cover worldwide—the equivalent of losing all stony corals on the Great Barrier Reef in Australia (Scouter et al., 2021a). In contrast, algae

---

1 The Global Coral Reef Monitoring Network was established by the International Coral Reef Initiative in 1995 to track and report coral reef status and trends. The 2020 report used a global dataset from 1978 to 2019, which comprised 2 million observations from 12,000 sites across 73 countries (Global Coral Reef Monitoring Network, 2020). The report uses live coral cover as the ecological parameter to assess the status of coral reefs. Calculating the percent of live coral cover present on a reef is one of the most frequently used parameters to measure coral reef health (Hill and Wilkinson, 2004).

cover on reefs, an indicator of stressed coral reefs, has increased by 20 percent in 2019 when compared to 2010, and the coral-to-algae ratio continues to increase in favor of algae. This is dire for the people who live next to reefs and depend on them for livelihoods. When people board an airplane and fly across the world to go diving, they do so for the coral species that are rapidly disappearing.

Caribbean Region

Carysfort Reef off of Key Largo in Florida, USA. Photograph by Sabine Bailey.

Coral cover and loss have varied across Earth's regions and time periods. The earliest observations of significant coral loss were documented in the Caribbean (Côté et al., 2005; Cramer et al., 2021). The Caribbean region represents ~7 percent of the world's reefs (Jackson et al., 2014). Some of the largest barrier reefs in the world are found in the Caribbean Sea. These include the Florida Keys Reef Tract which stretches 360 miles along Florida's coast in the United States as well as the Mesoamerican Reef which is nearly 700 miles long, spanning the coasts of Belize, Guatemala, Honduras and Mexico.

Average coral cover, which varies greatly across locations in the Caribbean, was estimated to be 14.3 percent in 2012 (Jackson et al., 2014). There has been a tragic loss of 50 to 80 percent in coral cover in the Caribbean since the 1970s (Côté et al., 2005; Jackson et al., 2014). Average coral cover declined from

34.8 percent in 1970 to 16.3 percent in 2011 (Jackson et al., 2014). A meta-analysis found that coral cover in the Caribbean decreased from 55 percent in 1977 to 13 percent in 2001, a colossal 80 percent decline in 24 years (Côté et al., 2005).

This loss has mainly been attributed to local stressors: (1) a mass die-off of long-spined sea urchins in the 1980s,[2] (2) followed by the widespread occurrence of coral disease, namely White Band Disease, which killed 80 percent of the remaining elkhorn coral (*Acropora palmata*), and (3) coral bleaching events in the 1990s (Cramer et al., 2021). These same threats of coral bleaching and coral disease continue to persist today (Alvarez-Filip, 2019; Walton et al., 2018). For instance, Stony Coral Tissue Loss Disease, a highly transmissible and deadly coral pathogen, which emerged in 2014, has contributed to up to a 30 percent additional loss in regional coral cover on the Florida Reef Tract (Walton et al., 2018). In 2018, Stony Coral Tissue Loss Disease spread quickly from Florida throughout the Caribbean in just a few short months, and is now present in Mexico, the U.S. Virgin Islands and the Dominican Republic (Alvarez-Filip, 2019).

Coral reefs in the Caribbean used to be dominated by impressive fields of stony coral colonies, with *Acropora* coral species, such as staghorn and elkhorn, reaching over five feet tall (Cramer et al., 2021; Dustan, 2003). However, the drastic decline in coral cover, which especially impacted *Acropora* coral species, has transformed reefs in the Caribbean into ecosystems dominated by algal communities and other opportunistic corals (Cramer et al., 2021; Jackson et al., 2014; Wilson et al., 2020). On average, algae cover has increased from 7 percent in 1984 to 23.6 percent in 1998 (Jackson et al., 2014). Algae-dominated reefs continue to persist to this day (Jackson et al., 2014; Wilson et al., 2020).

Researchers have also observed a change in the dominant coral species on reefs. Coral communities in the Caribbean have shifted from dominant and competitive branching corals, such as staghorn and elkhorn to stress-tolerant and weedy corals, such as star boulder coral (Cramer et al., 2021).[3]

2 Sea urchins are prominent herbivores that graze on algae, helping maintain the balance between algae cover and coral cover on reefs (Cramer et al., 2020).

3 Stress-tolerant corals are characterized by their slow to moderate growth, boulder-like morphologies, and higher thermal and sedimentation tolerance, such as the boulder star coral. Species include *Colpophyllia natans*, *Pseudodiploria* spp., *Meandrina* spp., *Montastrea cavernosa*, *Orbicella* spp., *Siderastrea* spp., and *Stephanocoenia* spp. Weedy corals are characterized by their low-relief branching and plating morphologies and high sexual recruitment and sedimentation tolerance and include *Agaricia* spp., *Madracis* spp., branching *Porites* spp., and *P. astreoides*. (Cramer et al., 2020).

These changes, known as ecological phase shifts, have shifted our baseline understanding of coral reefs in the Caribbean dramatically. What we once thought of as a coral reef now looks completely different. Current and future generations will never see and experience the endless fields of coral reefs that existed a mere 30 years ago. While the Caribbean may have been the first to witness such catastrophic coral loss, similar losses are currently unfolding in different regions around the world.

Australian Region

The most well-known and iconic coral reef, the Great Barrier Reef in Australia, has recently experienced severe coral losses. The Great Barrier Reef is the longest barrier reef in the world which stretches 1,429 miles along the eastern coast of Australia, covers an enormous area of 133,000 square miles, and is a World Heritage Site (Scouter et al., 2021b). The Australian region represents just over 16 percent of the world's coral reefs. The estimated live coral cover in 1994 was 35.2 percent and has declined to 26 percent in 2019, representing a 55 percent decrease (De'Ath et al., 2012; Scouter et al., 2021b). Over that time period, the Great Barrier Reef saw fluctuations of coral losses as well as remarkable recovery, demonstrating the resilience of these reefs, or their capacity to absorb stressors and return to their previous state. In the last 15 years, however, there has been a general trend of decreasing coral cover, which persists today (De'Ath et al., 2012; Fine et al., 2019a; Scouter et al., 2021b). Climate change is the biggest threat to the Great Barrier Reef, including both coral bleaching and ocean acidification impacts (Hughes et al., 2017b; Pendleton et al., 2019). 2016 marked the year of the most severe and widespread mass coral bleaching event in the Great Barrier Reef, bleaching 60 percent of corals (Hughes et al., 2017b). Other local stressors, such as population outbreaks of crown-of-thorns starfish, nuisance species which devour large portions of stony coral, have significantly impacted the loss of coral cover across the Great Barrier Reef (Westcott et al., 2020).

Similarly to the Caribbean, there has been an increase of algae cover as well as an emerging shift in dominant coral species composition (Fine et al., 2019a; Scouter et al., 2021b). Average algae cover was 21.2 percent in 1994 but has since increased to 37.5 percent in 2019, indicating a 77 percent increase in that time period (Scouter et al., 2021b). Coral species' abundance, diversity and size have declined, most prominently in the northern and central areas of the Great Barrier Reef (Dietzel et al., 2020; Fine et al., 2019a). Major shifts, with little indication in returning, in coral species composition have also been observed in the northern regions. Massive coral species and soft coral species are replacing branching coral species, reducing the structural complexity of the reef (Fine et al., 2019a). The Great Barrier Reef is iconic, due to its massive size, area and prominence. However, we are in the midst of witnessing large, disturbing and irreparable cracks in its foundation.

East Asian Region

The East Asian region encompasses coral reefs in East and Southeast Asia within the Pacific and Indian oceans (Scouter et al., 2021c). This region is home to the Coral Triangle, a marine ecosystem which is recognized as the epicenter of marine biodiversity, defined as the variety of life on Earth, and whose waters include those of Indonesia, Malaysia, the Philippines, Papua New Guinea, Timor-Leste and Solomon Islands (Asaad et al., 2018; Scouter et al., 2021c). The Coral Triangle contains over 76 percent of all coral species (approximately 600 species), over 2000 reef fish species, six out of seven sea turtle species, and has the largest mangrove forest in the world, a forest ecosystem connecting with seagrass and coral reefs makes up an entire ecosystem known as the tropical coastal seascape (Asaad et al., 2018; Scouter et al., 2021c; Veron et al., 2009). It directly supports the foundations of economically important pelagic (or open ocean) fisheries, such as tuna, as well as the livelihoods of over 120 million of people in the region (Cabral et al., 2012; Cros et al., 2014; Scouter et al., 2021c). The East Asian Region represents a significant portion of the world's reefs, housing just over a third (~30%) of all of the world's reefs (Scouter et al., 2021c).

The average coral cover was estimated to be 32.8 percent in 1983 and increased to 36.8 percent in 2019, with rising and falling fluctuations of percent coral cover over those 37 years (Scouter et al., 2021c). Despite the overall increase in coral cover, in the last decade coral cover has been declining, with a loss of 10.5 percent; with 90 percent of this loss occurring between 2005 and 2015 (Scouter et al., 2021c). However, an absence of data has been cited with regard to coral cover data in the East Asian region, and some researchers found that coral cover was actually lower than expected at ~22 percent in 2003 (Bruno & Selig, 2007; Scouter et al., 2021c). Due to the scarcity of data, future trends and algae cover in the East Asian Region are difficult to predict (Scouter et al., 2021c). It has been estimated that more than 85 percent of reefs in the Coral Triangle are threatened, which is considerably higher than the global average of 60 percent (Burke et al., 2012; Cabral et al., 2012). Local stressors, such as inland and marine pollution, unsustainable and destructive fishing practices and coastal development, are the primary threats. Climate change and the associated warming sea surface temperatures and ocean acidification impacts exacerbate stressors on the reefs (Burke et al., 2012; Cabral et al., 2012). As such, it is imperative to gather more data and mitigate the local and global threats to prevent unrestorable reef phase shifts as we have witnessed in the Caribbean and the Great Barrier Reef (Cramer et al., 2021; Fine et al., 2019a; Lowe et al., 2011).

Pacific Region

Ulong Channel Reef in Palau. Photograph by Nik Varley.

The Pacific region covers a vast area that comprises over 25,000 islands with reefs varying greatly in structure, complexity, biodiversity and threats (Lebrec et al., 2019; Scouter et al., 2021d). The Pacific region supports approximately 27 percent of the world's coral reefs (Scouter et al., 2021d). Some countries, states and territories in this region include Hawaii, Fiji, Palau, Papua New Guinea and French Polynesia. Coral reefs are integral toward providing a way of life, sustenance and economic opportunities via fisheries, dive and snorkel tourism to the majority of the people living on these islands (Lebrec et al., 2019; Scouter et al., 2021d). Population density also varies greatly across the region. As such, reefs in some areas are undisturbed by anthropogenic threats while others are severely degraded. This allows us a unique glimpse into the past and an opportunity to generate historic baselines which reflect a time prior to human disturbances (Smith et al., 2016).

Average coral cover in the Pacific region has remained largely stable at ~37 percent until 1998, where we observed a general decline to 31 percent stony coral cover in 2019 (Scouter et al., 2021d). Periods of recovery occurred during this time frame as well as periods of considerable decline, especially during El Niño years in 2015 and 2016. The majority of coral cover loss (~90 percent) transpired in the last decade, suggesting an accelerating trend toward further decline (Scouter et al., 2021d). By contrast, averages of algae cover of the seafloor on the coral reef habitat have progressively increased in the last two decades. Algae cover was stable at ~15 percent until 1999, and since then has increased to ~21 percent in 2018, with further evidence that it will continue to increase (Scouter et al., 2021d). The most imminent threat to Pacific coral reefs is climate change, which will increase coral bleaching events and ocean acidification (Lebrec et al., 2019; Scouter et al., 2021d). As such, managing and monitoring our global emissions are crucial for these yet increasingly vulnerable reefs, some of which are still in pristine condition (Smith et al., 2016).

Red Sea Region

The Red Sea is a unique body of water and is the second most biologically diverse coral reef community after Southeast Asia's Coral Triangle (Berumen et al., 2013; Fine et al., 2019b; Scouter et al., 2021e). The Red Sea contains over 1,000 fish species and over 340 scleractinian coral species (Berumen et al., 2013; DiBattista et al., 2016; Fine et al., 2019b). The length of the coral system in the Red Sea is remarkably almost as long as the Great Barrier Reef, at 1,242 miles, though only half of its area (Berumen et al., 2013). The Red Sea is unique due to its isolated fauna, high water temperature (reaching upward of 35°C in the summer) and high salinity (Berumen et al., 2013; DiBattista et al., 2016). The Red Sea is bordered by nine countries, including Djibouti, Egypt, Eritrea, Israel, Jordan, Saudi Arabia, Somalia, Sudan and Yemen. It represents just over 5 percent of the world's coral reefs (Scouter et al., 2021e).

Average coral cover in the Red Sea was estimated to be 36 percent in 1997 and has remained relatively stable (Scouter et al., 2021e). The Red Sea reefs have seen periods of recovery and decline corresponding to global mass coral bleaching events in 1998 and 2015–2016, where coral cover declined to 30.9 percent (Scouter et al., 2021e). However, since then, average coral cover recovered to approximately 34 percent in 2019. Average algae cover remained relatively stable from 1997 to 2012 at ~8 percent. However, algae cover almost doubled to 14.7 percent between 2012 to 2019, mirroring the same recent acceleration patterns that we have seen in other regions (Scouter et al., 2021e).

It is important to note that compared to coral reef research in the Caribbean and the Great Barrier Reef, the Red Sea coral reefs have had 1/6th and 1/8th of the amount of research conducted, respectively (Berumen et al., 2013). Furthemore, over half of the research has been concentrated in the northern part of the Red Sea which represents <2 percent of the region (Berumen et al., 2013; Scouter et al., 2021e). Finally, a large majority of sites in the region have also only been surveyed once (Scouter et al., 2021e). However, due to the unusual environmental conditions in the Red Sea (e.g., high water temperatures, high salinity), there is increasing attention from scientists to study the reefs in this region to help predict how reefs might fare in future warming and extreme conditions (Berumen et al., 2013).

While the health of coral reefs in the Red Sea seems relatively stable, geography-dependent local threats such as coastal development in the northern area, eutrophication in the central area and overfishing in the southern area are ever present (Berumen et al., 2013; Fine et al., 2019b; Scouter et al., 2021e). Climate change exacerbates local threats, alongside a rapidly growing population, which will likely place greater pressure on coastal development (Fine et al., 2019b). As such, increasing research and careful management will be critical to ensure the health of these unique reefs.

Overview

Overall, there is a general decline in global coral cover; over 50 percent of all hard corals have been lost in the last 30 years, exacerbated by imminent anthropogenic global and local threats (Eddy et al., 2021; Hoegh-Guldberg et al., 2019; Scouter et al., 2021a). The Caribbean and the Great Barrier Reef have or are experiencing major ecological phase shifts from branching coral-dominated reefs to algal-dominated ones (Cramer et al., 2021; Fine et al., 2019a; Lowe et al., 2011).

Despite the grim outlook, there is still hope for our reefs. Under the right conditions, recovery is possible and has been observed after mass coral bleaching events. In these instances, global average coral cover increased between 2002 and 2009, and in 2019. Recently, it was found that widespread coral recovery was taking place in the Great Barrier Reef after only a year of no major disturbances (e.g., coral bleaching event, typhoons, oil spills) (Australian Institute of Marine Science, 2021). The key is to mitigate climate change impacts and as many disturbances and threats as possible, while considering where the coral reefs are located, the people who depend on them, the governing structures in place and the localized pressures (Australian Institute of Marine Science, 2021; Scouter et al., 2021a).

This means that management should become regionally distinct, requiring the input and buy-in of the coastal people that depend on the reefs. This form of participatory management must also be flexible and adaptive to protect these vulnerable and brilliant ecosystems.

Chapter 4

APPROACHES TO CORAL REEF MANAGEMENT

By Amanda Alva, Daniel Morris, Kelly Dunning

In the face of rapid global environmental change, there has been an interdisciplinary movement among scientists to reconsider how ecosystems are managed. Governing in the Anthropocene, or the age in which human activity has become the dominant influence on the Earth's climate and environment, comes with novel considerations. As the human population continues to grow, interactions between people and the environment are becoming more common and more complicated. These interactions are often described as socio-ecological systems, or the relationships between linked human and natural systems (Folke, 2006; Folke et al., 2005; Ostrom, 2009; Walker et al., 2004).

Due to the fact that coastal communities rely on coral reef ecosystems, the relationships between them are important to consider as a single socio-ecological system (Cinner et al., 2012; Cinner et al., 2016; Kittinger et al., 2016). This means that impacts to the reef, through human action or natural change, can come back to impact human systems like economies and cultures. For example, fishermen access a reef for different species, such as conch, lobster, snapper and grouper. These different anglers might overlap in their target areas, limiting the ability of other reef users, such as divers, to view these species. People looking to develop stretches of the coastline for condos may clear coastal dunes, resulting in cloudier water that inhibits both angler and diver activity. A member of the public that may never even see a reef firsthand might place a high value on a healthy reef merely existing so that they can be passed down to their children. Decision-makers are tasked with understanding these overlapping uses, and designing management systems to balance uses while maintaining essential features of the reef ecosystem.

Dr. Elinor Ostrom, a pioneer of the socio-ecological system concept, likens the complexity of a resource system to the complexity of a biological organism, where many subsystems interact to produce feedback loops which compose a greater whole (Ostrom, 2009). While the idea of a socio-ecological system was initially developed to help explain the interconnected nature of people and their environment, it has since become a staple within the field of natural resources management being applied to topics as diverse as cattle ranching, forestry and coral reef conservation (Berkes et al., 2000; Fischer, 2018; Herrero-Jáuregui et al., 2018; Pendleton et al., 2016; Walker et al., 2004).

The pressures facing coral reefs and the impacts to social and economic systems vary depending on where reefs are located (Cinner et al., 2005). In lower-income, developing nations where the practice of subsistence fishing is higher, due to higher numbers of people facing a lack of food security, destructive fishing methods, such as blast fishing, are more common (Hampton-Smith et al., 2021). Blast fishing is a process where anglers mix inexpensive household chemicals to form homemade "bombs," which stun larger species of fish that float to the top to be collected and sold at the market by fishermen. In wealthier countries, unchecked coastal development can impact reefs just as seriously, creating water clarity and water quality issues in coastal waters that may prevent reefs from thriving.

Understanding the dynamics of socio-ecological systems is critical in both emerging markets and high-income countries. For example, among all of the 29 World Heritage-Listed Coral Reefs, including the Great Barrier Reef in Australia, the most commonly reported threat was poor reef management and governance, a problem spanning both developed and lesser developed countries (Morrison et al., 2020). Even in developed countries which have more ample resources and enforcement capabilities for rules protecting reefs, managing natural resources is a challenge, especially when the livelihoods of so many stakeholders depend on them.

Given the challenges of understanding and then managing socio-ecological systems and high levels of variation according to regions, nation-states and even communities adjacent to reefs, unique management strategies are necessary. Management that seeks out stakeholder involvement in decisions, also known as *participatory management*, ensures the complexity of socio-ecological systems can be accounted for.

The Evolution of Participatory Coral Reef Management

In the West, in general, coral reefs are managed by a bureaucracy, where the government, in the form of a resource agency, sets the rules and regulations on behalf of the public, usually amid markets for activities like dive tourism and commercial fishing. Bureaucracies set the terms in the form of regulations, governing resource users (Imperial & Yandle, 2005). In the United States, federal agencies like the National Oceanic and Atmospheric Administration or NOAA, state fish and wildlife agencies, and local municipalities set different rules governing access to American reefs, with significant variation expanded on in Chapters 5 and 9. Bureaucracies may sometimes fail in their efforts aimed at conservation due to a lack of (1) resources, (2) understanding of local needs, or (3) capacity to enforce rules. Bureaucracy may be so large that it fails to incorporate local knowledge of people that use the reef. Bureaucracy may impose regulations that result in depleted resources and decreased livelihoods (Ayers & Kittinger, 2014). That is not always the case, but these flaws are the impetus for increasing participation in institutions.

Contrasting with the bureaucracy, *community-based resource management* institutions received a significant increase in attention from resource managers and decision-makers looking to conserve coral reefs in a participatory way (Johannes, 2002; Pomeroy, 1995). Community-based resource management is where day-to-day resources users, or stakeholders, such as fishermen, develop their own rules to regulate the resource among themselves. Generally, community-based resource management must have the following characteristics to be successful: (1) strong leadership and community initiative/desire to implement the program is present (Ayers & Kittinger, 2014; Dunning, 2021; Haley & Clayton, 2003; Imperial & Yandle, 2005); (2) regulation of resources is reflective of cultural and political context (Blaikie, 2006; Cinner et al., 2005; White & Vogt, 2000); and (3) interpersonal trust, an interest in learning, and ways to reach shared understanding within communities are present (Ayers & Kittinger, 2014; Cinner et al., 2005; Christie et al., 1994; Imperial & Yandle, 2005). Without these components, conflicts may destroy the institution.

A third type of natural resource governance is called *co-management*, which can be thought of as a combination between bureaucracy and community-based institutions (Ayers & Kittinger, 2014). Co-management is a hybrid institutional arrangement in which both the government and stakeholders share the responsibility of managing the natural resources (Imperial, 2008). This arrangement includes the government effectively assigning property rights to a resource and partnering with stakeholders to enable them to make decisions on the resource over which they hold a form of tenure. An example of co-management in the developed world can be found in Hawaii, where subsistence fishing areas have been developed by the government to enable local communities to propose management ideas for ocean resources (Ayers & Kittinger, 2014).

Success of co-management has been varied, where it's been found to possibly inflame conflicts between competing use groups (Gelcich et al., 2006) while also being beneficial to socio-ecological systems improving the resource and socioeconomics (Cinner et al., 2012). Co-management is successful when there is trust between the government and communities, well-developed understanding of responsibilities and roles, equitable distribution of power and management and incorporation of local knowledge (Ayers & Kittinger, 2014; Cinner et al., 2012; Clarke & Jupiter, 2010; Gelcich et al., 2006).

Both community-based resource management and co-management are participatory in their nature. Participatory management of coral reefs that includes stakeholders to make decisions and manage the resource, has been increasing since the 1970s. First emerging in Southeast Asia, where the need to include resource users was first realized and implemented by governments and multilateral aid donors, the practice now occurs all over the world (Amar et al., 1996; Christie et al., 1994; Pomeroy, 1995; White & Vogt, 2000). Small-scale reef fisheries, which provide substantial portions of many low-income countries' nutrition, were facing a decline. This decline and the accompanying food insecurity set these communities into action (Amar et al., 1996; White & Vogt, 2000). Initially, fishermen were under the impression that management of their coral reef resources was beyond their control, better suited for their government, which is a distant, top-down manager of their important food sources. Fishermen blamed their government for the deteriorating reef ecosystems (Christie et al., 1994). An alternative model was implemented by community leaders, development experts and decision-makers. Participatory management was seen as a pathway to reduce conflicts among users and create more equitable and effective resource distribution throughout communities. Participatory managers aimed to enact a system where wealthy and low-income people alike had suitable access to fisheries and improved food security (Pomeroy, 1995).

More traditional, well-known institutions for coral reef management were also being applied on a wider scale beginning in this same time frame. These institutions include agency-based decision-making, enacting laws and implementing effective law enforcement to govern reef access to fisheries and tourism (dive sites); end destructive fishing; establish MPAs; and restrict the harvest of certain protected species like ornamental reef fish (Amar et al., 1996; McClanahan et al., 2006; White & Vogt, 2000). Several examples of stakeholder inclusion in management include the Community Fishery Resource Management Project in the Philippines and the Marine Conservation Project for San Salvador, located in the village of Masinloc municipality in Zambales, Philippines. These two programs began to expand traditional institutions for coral reef management to engage and include users in a participatory fashion (Amar et al., 1996; Christie et al., 1994; Pomeroy, 1995; White & Vogt, 2000).

Understanding real-world examples of participatory reef management is essential to mimicking their successes and avoiding pitfalls worldwide. For example, the Community Fishery Resource Management Project in the Philippines involved culturally traditional management institutions on Malalison Island, such as the local legislative assembly (*Culasi Sangguniang Bayan*), fishermen's associations, traditional national governmental agencies and non-governmental organizations to work with fishermen to co-manage the coral reef fisheries. Actions included the creation of a fish sanctuary and deploying artificial reefs made of concrete (Agbayani & Babol, 2001). The fish sanctuary assists with localized pressures such as overfishing with artificial reef installation acting as a response both to localized pressures as well as larger-scale climate pressures.

The Marine Conservation Project for San Salvador Island, for example, was put in place to organize certain areas on the reef where fishing was prohibited altogether, as well as certain areas where traditional fishing gear and practices were the only permitted forms of fishing. A community organization was formed and tasked with decision-making and enforcement of this participatory management institution (Christie et al., 1994). Indigenous expertise in fisheries management was leveraged to enact this co-managed protected area.

In Indonesian communities on the island of Bali, communities like Pemuteran, Amed and Lovina have complex local institutions to govern their MPAs. Lovina has associations of fishermen and dolphin tour guide operators who enact rules to govern access to reefs and control pressure placed on dolphins, limiting the number of boats on the water. These limitations are self-initiated in the name of marine mammal protection. The village of Pemuteran has its own undersea police force, known as the *pecelan laut*, a culturally and religiously relevant institution based on local ways of knowing. These police wear traditional clothing and engage tourists on the beach, acting as both educational guides and law enforcement. Pemuteran also has special diving zones where tourists pay U.S. $2 more as a conservation fee to visit. These funds go to train youths to act as conservation practitioners in community programs. Amed has community-created artificial reefs, placed on the seafloor during local cultural ceremonies, that act as alternative dive sites to relieve some pressure on the reef (Dunning, 2018, 2022).

In another example of participatory coral reef management, from Kenya and Madagascar, communities were recognized for having MPAs with clearly defined membership rights where reef users contributed to the management and could access the reefs for livelihoods. Conflict resolution mechanisms, such as procedures for what to do when someone breaks the rules, and a set of rules governing the MPA that matched up with local conditions and needs, were core components of management (Cinner et al., 2009, p. 493). One important aspect of participatory management is successful enforcement of rules designed with consultation from nearby communities, enforced by people from the community, factors that can help increase both income from fisheries and food yields (White & Vogt, 2000).

Adapting to Environmental Change: The Need for Adaptive Governance

Participatory management of coral reefs affords more opportunities for stakeholders to be included in management. These management institutions must also be capable of changing as rapidly as possible due to the stressors impacting coral reef environments. One form of management that enables a timely response to environmental change is that of *adaptive governance.* Adaptive governance is composed of regulations, processes and organizations aimed at achieving desired environmental outcomes, or in our case, reefs with functional ecosystems (Lemos & Agrawal, 2006).

Adaptive governance, at its core, is about decision-makers learning to respond to environmental disturbances in ways that build resilience. In other words, adaptive governance focuses on learning-based approaches to managing ecosystems, where knowledge and understanding are used over time to improve management practices across multiple institutional, governmental, spatial and temporal scales, enabling a socio-ecological system to overcome rapid, unpredictable changes (Folke, 2005; Schultz et al., 2015).

Adaptive governance is a departure from traditional, bureaucratic governance in that it relies heavily on participatory features, like stakeholder involvement in decision-making (Brunner et al., 2005). Worldwide, adaptive governance has been used to design and assess many coral reef management interventions such as MPAs (Ban et al., 2011; Evans et al., 2011) and co-management strategies (Cinner et al., 2009; McClanahan et al., 2006). Adaptive governance has been used as a tool to assess the Great Barrier Reef (McCook et al., 2010; Morrison, 2017).

A recent literature review of adaptive governance highlights a mismatch between different types of institutions and their approaches to conservation governance when managing socio-ecological systems. For example, centralized governance often fails when working within large-scale ecosystems due to jurisdictional overlap, whereas bottom-up collective approaches to governance, such as polycentric governance, may work better in these instances (Chaffin et al., 2014). This is also termed "institutional-fit," where adaptive governance is used to assess or improve institutions and their modes of governance to best fit the scenario and achieve desirable environmental outcomes (Lockwood et al., 2012; Rijke et al., 2012; Sternlieb et al., 2013).

Governance becomes adaptive (or governance that enables learning by doing) when decision-makers enact conservation interventions for an ecosystem and learn from the results that occur over time, thereby reducing uncertainty for decision-makers for their next management intervention. Dietz et al. argue that three social conditions are necessary to ensure adaptive governance: (1) inclusive dialogue between stakeholders, (2) the involvement of all necessary stakeholders (e.g., government at all scales, private sector, and community groups), and (3) the ability for decision-makers to experiment and learn.

An example of adaptive governance is occurring in response to the immense environmental change borne by the Florida Reef Tract. A wide array of actors, from inside and outside of formal government, are responding to these changes. For example, federal government managers in the Florida Keys National Marine Sanctuary and the NOAA Office of Habitat Conservation are working with several non-governmental organizations, universities, marine research institutions and the private sector to restore 3 million square feet of the Florida reef tract. They are doing so using the best available science, including replanting corals that are heat tolerant and disease resistant. Corals that have taken hundreds of experts and volunteers alike thousands of hours to grow in nurseries and replant. Interviews with coral reef restoration professionals, specifically a company called Reef Renewal USA, highlight novel approaches to reef restoration that emphasize rapid response, rapid scaling and innovation:

> *I am working to restore coral reefs as quickly as possible because I am a father and a local. I want the same reefs that I had growing up to be there for my children. We are completely open source. We post and share all of our how-to guides online for free. We are not guarding our technology for growing and replanting corals as fast as possible. We are trying to innovate quickly and spread the word just as fast. Operating in the private sector helps us move faster. We also partner with educational organizations that provide outreach to young folks and the larger community on how we restore corals. We believe that by effectively engaging the next generation of reef restoration volunteers, we create the greatest hope for the future for coral reef restoration.*

One such educational and outreach example is the "coral tree" nursery structure depicted in Figure 4.1. This tree is intended as an exhibit to educate the public on coral restoration, located at Aquarium Encounters in Marathon, Florida. The technology of the coral tree is a major innovation in restoration, enabling more favorable hydrological conditions as well as access to sunlight.

Figure 4.1 Coral tree at Aquarium Encounters in Marathon, Florida. Source: Image courtesy of Ben Daughtry.

Trees are attached to the ocean floor, buoyed with a float and are able to move with the water to survive large waves during a hurricane.

At no point in our history have so many organizations of various types (government, non-governmental organizations and the private sector) worked at such a large scale to meet such an urgent challenge. Our book provides several more novel insights regarding adaptive governance and participatory management of coral reefs, where local communities are invited to co-manage reef systems. We present five case studies showing stakeholder-driven policy-making at the national level through bipartisan legislation in the United States, at the sub-national level through participatory management in Florida, and at the local level through grassroots movements in Florida and the Cayman Islands.

REFERENCES PART 1

Agbayani, R., & Babol, A. (2001). Community-based Fishery Resources Management Project in Malalison Island: Institutional arrangements for fisheries co-management. SEAFDEC/AQD Institutional Repository. Retrieved January 2021, from https://repository.seafdec.org.ph/handle/10862/474

Albright, R., Mason, B., Miller, M., & Langdon, C. (2010). Ocean acidification compromises recruitment success of the threatened Caribbean coral Acropora palmata. *Proceedings of the National Academy of Sciences*, *107*(47), 20400–20404.

Alvarez-Filip, L., Estrada-Saldívar, N., Pérez-Cervantes, E., Molina-Hernández, A., & González-Barrios, F. J. (2019). A rapid spread of the stony coral tissue loss disease outbreak in the Mexican Caribbean. *PeerJ*, *7*, e8069.

Amar, E. C., Cheong, R. M. T., & Cheong, M. V. T. (1996). Small-scale fisheries of coral reefs and the need for community-based resource management in Malalison Island, Philippines. *Fisheries Research*, *25*(3–4), 265–277. https://doi.org/10.1016/0165-7836(95)00439-4

Andrello, M., Darling, E., Wenger, A., Suarez-Castro, A. F., Gelfand, S., & Ahmadia, G. (2021). A global map of human pressures on tropical coral reefs. *bioRxiv*, *15*(1).

Asaad, I., Lundquist, C. J., Erdmann, M. V., & Costello, M. J. (2018). Delineating priority areas for marine biodiversity conservation in the Coral Triangle. *Biological Conservation*, *222*, 198–211. https://doi.org/10.1016/j.biocon.2018.03.037

Australian Institute of Marine Science. (2021). *Background Reef in recovery window after decade of disturbances*. https://www.aims.gov.au/sites/default/files/2021-07/AIMS%20Long%20Term%20Monitoring%20Program%20Annual%20Summary%20Report%20on%20Coral%20Reef%20Condition%20%28GBR%29%202020-2021_July2021.pdf

Ayers, A. L., & Kittinger, J. N. (2014). Emergence of co-management governance for Hawai'i coral reef fisheries. *Global Environmental Change*, *28*(1), 251–262. https://doi.org/10.1016/j.gloenvcha.2014.07.006

Ban, N. C., Adams, V. M., Almany, G. R., Ban, S., Cinner, J. E., McCook, L. J., Mills, M., Pressey, R. L., & White, A. (2011). Designing, implementing and managing marine protected areas: Emerging trends and opportunities for coral reef nations. *Journal of Experimental Marine Biology and Ecology*, *408*(1–2), 21–31.

Beger, M., Harborne, A. R., Dacles, T. P., Solandt, J. L., & Ledesma, G. L. (2004). A framework of lessons learned from community-based marine reserves and its effectiveness in guiding a new coastal management initiative in the Philippines. *Environmental Management*, *34*(6), 786–801.

Berkes, F. (2010). Devolution of environment and resources governance: Trends and future. *Environmental Conservation*, *37*(4), 489–500.

Berkes, F., Folke, C., & Colding, J. (Eds.). (2000). *Linking social and ecological systems: Management practices and social mechanisms for building resilience.* Cambridge University Press.

Berkes, F., Colding, J., & Folke, C. (Eds.). (2008). *Navigating social-ecological systems: Building resilience for complexity and change.* Cambridge University Press.

Berumen, M. L., Hoey, A. S., Bass, W. H., Bouwmeester, J., Catania, D., Cochran, J. E. M., Khalil, M. T., Miyake, S., Mughal, M. R., Spaet, J. L. Y., & Saenz-Agudelo, P. (2013). The status of coral reef ecology research in the Red Sea. In *Coral reefs* (Vol. 32, No. 3, pp. 737–748). Springer. https://doi.org/10.1007/s00338-013-1055-8

Blaikie, P. (2006). Is small really beautiful? Community-based natural resource management in Malawi and Botswana. *World Development, 34*(11), 1942–1957. https://doi.org/10.1016/j.worlddev.2005.11.023

Brunner, R. D., Steelman, T. A., Coe-Juell, L., Cromley, C. M., Edwards, C. M., & Tucker, D. W. (2005). *Adaptive governance: Integrating science, policy, and decision making.* Columbia University Press.

Bruno, J. F., & Selig, E. R. (2007). Regional decline of coral cover in the Indo-Pacific: Timing, extent, and subregional comparisons. *PLoS ONE, 2*(8), e711. https://doi.org/10.1371/journal.pone.0000711

Burke, L., Reytar, K., Spalding, M., & Perry, A. (2012). *Reefs at risk revisited.* World Resources Institute.

Cabral, R., Cruz-Trinidad, A., Geronimo, R., & Aliño, P. (2012). Opportunities and challenges in the coral triangle. In *Environmental science and technology* (Vol. 46, No. 15, pp. 7930–7931). American Chemical Society. https://doi.org/10.1021/es3025462

Chaffin, B. C., Gosnell, H., & Cosens, B. A. (2014). A decade of adaptive governance scholarship: Synthesis and future directions. *Ecology and Society, 19*(3). https://www.jstor.org/stable/26269646

Christie, P., White, A. T., & Buhat, D. (1994). Community-based coral reef management on San Salvador Island, the Philippines. *Society and Natural Resources, 7*(2), 103–117. https://doi.org/10.1080/08941929409380849

Cinner, J., & Huchery, C. (2014). A comparison of social outcomes associated with different fisheries co-management institutions. *Conservation Letters, 7*(3), 224–232.

Cinner, J. E., & Aswani, S. (2007). Integrating customary management into marine conservation. *Biological Conservation, 140*(3), 201–216.

Cinner, J. E., Marnane, M., McClanahan, T. R., Clark, T., & Ben, J. (2005). Trade, tenure, and tradition: Influence of sociocultural factors on resource use in Melanesia. *Conservation Biology, 19*(2), 504–511. https://doi.org/10.1111/j.1523-1739.2005.00101.x

Cinner, J. E., McClanahan, T. R., Daw, T. M., Graham, N. A., Maina, J., Wilson, S. K., et al. (2009). Linking social and ecological systems to sustain coral reef fisheries. *Current Biology, 19*, 206–212. https://doi.org/10.1016/j.cub.2008.11.055

Cinner, J. E., McClanahan, T. R., MacNeil, M. A., Graham, N. A., Daw, T. M., Mukminin, A., Feary, D. A., Rabearisoa, A. L., Wamukota, A., Jiddawi, N., Campbell, S. J., Baird, A. H., Januchowski-Hartley, F. A., Hamed, S., Lahari, R., Morove, T., & Kuange, J. (2012). Comanagement of coral reef social-ecological systems. *Proceedings of the National Academy of Sciences, 109*(14), 5219–5222.

Cinner, J. E., Pratchett, M. S., Graham, N. A. J., Messmer, V., Fuentes, M. M. P. B., Ainsworth, T., Ban, N., Bay, L. K., Blythe, J., Dissard, D., & Dunn, S. (2016). A framework for understanding climate change impacts on coral reef social–ecological systems. *Regional Environmental Change, 16*(4), 1133–1146.

Clarke, P., & Jupiter, S. D. (2010). Law, custom and community-based natural resource management in Kubulau District (Fiji). *Environmental Conservation*, *37*(1), 98–106. https://doi.org/10.1017/S0376892910000354

Conservation International. (2008). Economic values of coral reefs, mangroves, and seagrasses: A global compilation 2008. In Smithsonian. www.communities.coastalvalues.org/

Coral reefs and climate change | IUCN. (2021). Retrieved January 31, 2022, from https://www.iucn.org/resources/issues-briefs/coral-reefs-and-climate-change

Côté, I. M., Gill, J. A., Gardner, T. A., & Watkinson, A. R. (2005). Measuring coral reef decline through meta-analyses. *Philosophical Transactions of the Royal Society B: Biological Sciences*, *360*(1454), 385–395. https://doi.org/10.1098/rstb.2004.1591

Cramer, K. L., Donovan, M. K., Jackson, J. B. C., Greenstein, B. J., Korpanty, C. A., Cook, G. M., & Pandolfi, J. M. (2021). The transformation of Caribbean coral communities since humans. *Ecology and Evolution*, *11*(15), 10098–10118. https://doi.org/10.1002/ECE3.7808

Creary, M., Alcolado, P., Coelho, V., Crabbe, J., Green, S., Geraldes, F., Ainsley, H., Hibbert, H., Jones, R. O., Jones-Smith, L. O., & Manfrino, C. A. (2008). Status of coral reefs in the northern Caribbean and western Atlantic GCRMN node in 2008.

Crawford, S. E. S., & Ostrom, E. (1995). A grammar of institutions. *American Political Science Review*, *89*(3), 582–600. https://doi.org/10.2307/2082975

Cros, A., Fatan, N. A., White, A., Teoh, S. J., Tan, S., Handayani, C., Huang, C., Peterson, N., Li, R. V., Siry, H. Y., Fitriana, R., Gove, J., Acoba, T., Knight, M., Acosta, R., Andrew, N., & Beare, D. (2014). The coral triangle Atlas: An integrated online spatial database system for improving coral reef management. *PLoS ONE*, *9*(6). https://doi.org/10.1371/JOURNAL.PONE.0096332

De'Ath, G., Fabricius, K. E., Sweatman, H., & Puotinen, M. (2012). The 27-year decline of coral cover on the Great Barrier Reef and its causes. *Proceedings of the National Academy of Sciences of the United States of America*, *109*(44), 17995–17999. https://doi.org/10.1073/pnas.1208909109

Dibattista, J. D., Roberts, M. B., Bouwmeester, J., Bowen, B. W., Coker, D. J., Lozano-Cortés, D. F., Howard Choat, J., Gaither, M. R., Hobbs, J. P. A., Khalil, M. T., Kochzius, M., Myers, R. F., Paulay, G., Robitzch, V. S. N., Saenz-Agudelo, P., Salas, E., Sinclair-Taylor, T. H., Toonen, R. J., Westneat, M. W., … Berumen, M. L. (2016). A review of contemporary patterns of endemism for shallow water reef fauna in the Red Sea. *Journal of Biogeography*, *43*(3), 423–439. https://doi.org/10.1111/jbi.12649

Dietzel, A., Bode, M., Connolly, S. R., & Hughes, T. P. (2020). Long-term shifts in the colony size structure of coral populations along the Great Barrier Reef. *Proceedings of the Royal Society B: Biological Sciences*, *287*(1936), 20201432. https://doi.org/10.1098/rspb.2020.1432

Dispensa, J. M., & Brulle, R. J. (2003). Media's social construction of environmental issues: Focus on global warming – A comparative study. *International Journal of Sociology and Social Policy*, *23*(10). https://doi.org/10.1108/01443330310790327

Dunning, K. (2021). Adaptive governance: The proposed port expansion in the Cayman Islands and its impacts to coral reefs. *Marine Policy*, *124*, 104248.

Dunning, K. (2022). Biodiversity conservation policy in megadiverse countries: Comparing policy systems for 2020 targets to inform management in the coming decades. *Journal of Environmental Management*, *302*, 113815.

Dunning, K. H. (2018). *Managing coral reefs: An ecological and institutional analysis of ecosystem services in Southeast Asia*. Anthem Press.

Dustan, P. (2003). Ecological perspective: The decline of Carysfort Reef, Key Largo, Florida 1975–2000. NOAA Technical Memo, NOS NCCOS (1), 37–43.

Eddy, T. D., Lam, V. W. Y., Reygondeau, G., Cisneros-Montemayor, A. M., Greer, K., Palomares, M. L. D., Bruno, J. F., Ota, Y., & Cheung, W. W. L. (2021). Global decline in capacity of coral reefs to provide ecosystem services. *One Earth*, *4*(9), 1278–1285. https://doi.org/10.1016/j.oneear.2021.08.016

Evans, L. S., Brown, K., & Allison, E. H. (2011). Factors influencing adaptive marine governance in a developing country context: A case study of southern Kenya. *Ecology and Society*, *16*(2). https://www.jstor.org/stable/26268893?casa_token=7qxxq7KB3EwAAAAA%3Ag5Z1CTFSXnGv8URt4EXbbXpDJQSKB4mNbvmm2TkE1oMbSRIu-ph2NlV9vlVntgMJTx6xFmdeliv6GMeBsKOT-uleJ1xGh9re5euSjynW2otkf3BsmVY

Ferse, S. C. A., Glaser, M., Neil, M., & Máñez, K. S. (2014). To cope or to sustain? Eroding long-term sustainability in an Indonesian coral reef fishery. *Regional Environmental Change*, *14*(6), 2053–2065.

Fine, M., Cinar, M., Voolstra, C. R., Safa, A., Rinkevich, B., Laffoley, D., Hilmi, N., & Allemand, D. (2019b). Coral reefs of the Red Sea — Challenges and potential solutions. *Regional Studies in Marine Science*, *25*, 100498. https://doi.org/10.1016/j.rsma.2018.100498

Fine, M., Hoegh-Guldberg, O., Meroz-Fine, E., & Dove, S. (2019a). Ecological changes over 90 years at Low Isles on the Great Barrier Reef. *Nature Communications*, *10*(1), 1–8. https://doi.org/10.1038/s41467-019-12431-y

Fischer, A. P. (2018). Forest landscapes as social-ecological systems and implications for management. *Landscape and Urban Planning*, *177*, 138–147. https://doi.org/10.1016/j.landurbplan.2018.05.001

Folke, C. (2006). Resilience: The emergence of a perspective for social–ecological systems analyses. *Global environmental change*, *16*(3), 253–267.

Folke, C., Hahn, T., Olsson, P., & Norberg, J. (2005). Adaptive governance of social-ecological systems. *Annual Review of Environment and Resources*, *30*(1), 441–473. https://doi.org/10.1146/annurev.energy.30.050504.144511

Forsman R. B. (2005). Life and death on the coral reef: An ecological perspective on scholarly publishing in the health sciences. *Journal of the Medical Library Association: JMLA*, *93*(1), 7–15.

Frys, C., Saint-Amand, A., Le Hénaff, M., Figueiredo, J., Kuba, A., Walker, B., Lambrechts, J., Vallaeys, V., Vincent, D., & Hanert, E. (2020). Fine-scale coral connectivity pathways in the Florida Reef Tract: Implications for conservation and restoration. *Frontiers in Marine Science*, *7*, 312. https://doi.org/10.3389/fmars.2020.00312

Gelcich, S., Edwards-Jones, G., Kaiser, M. J., & Castilla, J. C. (2006). Co-management policy can reduce resilience in traditionally managed marine ecosystems. *Ecosystems*, *9*(6), 951–966. https://doi.org/10.1007/s10021-005-0007-8

Graham, N. A., Cinner, J. E., Norström, A. V., & Nyström, M. (2014). Coral reefs as novel ecosystems: Embracing new futures. *Current Opinion in Environmental Sustainability*, *7*, 9–14. https://doi.org/10.1016/j.cosust.2013.11.023

Gruber, N., Clement, D., Carter, B. R., Feely, R. A., Van Heuven, S., Hoppema, M., Ishii, M., Key, R. M., Kozyr, A., Lauvset, S. K., & Lo Monaco, C. (2019). The oceanic sink for anthropogenic CO2 from 1994 to 2007. *Science*, *363*(6432), 1193–1199.

Hampton-Smith, M., Bower, D. S., & Mika, S. (2021). A review of the current global status of blast fishing: Causes, implications and solutions. *Biological Conservation, 262*, 109307.

Haley, M., & Clayton, A. (2003). The role of NGOs in environmental policy failures in a developing country: The mismanagement of Jamaica's coral reefs. *Environmental Values, 12*(1), 29–54.

Henkel, T. P. (2010) Coral reefs. *Nature Education Knowledge, 3*(10), 12. https://www.nature.com/scitable/knowledge/library/coral-reefs-15786954/

Herrero-Jáuregui, C., Arnaiz-Schmitz, C., Reyes, M. F., Telesnicki, M., Agramonte, I., Easdale, M. H., Schmitz, M. F., Aguiar, M., Gómez-Sal, A., & Montes, C. (2018). What do we talk about when we talk about social-ecological systems? A literature review. *Sustainability, 10*(8), 2950.

Hicks, C. C., & Cinner, J. E. (2014). Social, institutional, and knowledge mechanisms mediate diverse ecosystem service benefits from coral reefs. *Proceedings of the National Academy of Sciences, 111*(50), 17791–17796.

Hill, J., & Wilkinson, C. (2004). Methods for ecological monitoring of coral reefs. Australian Institute of Marine Science, Townsville, 117.

Hoegh-Guldberg, O. (1999). Climate change, coral bleaching and the future of the world's coral reefs. *Marine and Freshwater Research, 50*, 839–866.

Hoegh-Guldberg, O., Pendleton, L., & Kaup, A. (2019). People and the changing nature of coral reefs. *Regional Studies in Marine Science, 30*, 100699.

Hoegh-Guldberg, O., Poloczanska, E. S., Skirving, W., & Dove, S. (2017). Coral reef ecosystems under climate change and ocean acidification. *Frontiers in Marine Science, 4*, 158.

Hopkin, M. (2005, November 24). Greenhouse-gas levels highest for 650,000 years. Retrieved January 20, 2022, from https://www.nature.com/news/2005/051121/full/051121-14.html

Hughes, T. P. (1994). Catastrophes, phase shifts, and large-scale degradation of a Caribbean coral reef. *Science, 265*(5178), 1547–1551.

Hughes, T. P., Anderson, K. D., Connolly, S. R., Heron, S. F., Kerry, J. T., Lough, J. M., et al. (2018b). Spatial and temporal patterns of mass bleaching of corals in the anthropocene. *Science, 359*, 80–83. https://doi.org/10.1126/science.aan8048

Hughes, T. P., Baird, A. H., Bellwood, D. R., Card, M., Connolly, S. R., Folke, C., Grosberg, R., Hoegh-Guldberg, O., Jackson, J. B. C., Kleypas, J., Lough, J. M., Marshall, P., Nyström, M., Palumbi, S. R., Pandolfi, J. M., Rosen, B., & Roughgarden, J. (2003). Climate change, human impacts, and the resilience of coral reefs. *Science, 301*(5635), 929–933. https://doi.org/10.1126/science.1085046

Hughes, T. P., Barnes, M. L., Bellwood, D. R., Cinner, J. E., Cumming, G. S., Jackson, J. B. C., Kleypas, J., van de Leemput, I. A., Lough, J. M., Morrison, T. H., Palumbi, S. R., van Nes, E. H., & Scheffer, M. (2017a). Coral reefs in the anthropocene. *Nature, 546*(7656), 82–90. https://doi.org/10.1038/nature22901

Hughes, T. P., Kerry, J. T., Alvarez-Noriega, M., Alvarez-Romero, J. G., Anderson, K. D., Baird, A. H., Babcock, R. C., Beger, M., Bellwood, D. R., Berkelmans, R., Bridge, T. C., Butler, I. R., Byrne, M., Cantin, N. E., Comeau, S., Connolly, S. R., Cumming, G. S., Dalton, S. J., Diaz-Pulido, G., … Wilson, S. K. (2017b). Global warming and recurrent mass bleaching of corals. *Nature, 543*(7645), 373–377. https://doi.org/10.1038/nature21707

Hughes, T. P., Kerry, J. T., Baird, A. H., Connolly, S. R., Dietzel, A., Eakin, C. M., Heron, S. F., Hoey, A. S., Hoogenboom, M. O., Liu, G., McWilliam, M. J., Pears, R.

J., Pratchett, M. S., Skirving, W. J., Stella, J. S., & Torda, G. (2018a). Global warming transforms coral reef assemblages. *Nature, 556*(7702), 492–496. https://doi.org/10.1038/s41586-018-0041-2

Imperial, M. T., & Yandle, T. (2005). Taking institutions seriously: Using the IAD framework to analyze fisheries policy. *Society and Natural Resources, 18*(6), 493–509. https://doi.org/10.1080/08941920590947922

Jackson, J., Donovan, M., Cramer, K., & Lam, V. (2014). Status and trends of Caribbean Coral reefs. In *Global Coral Reef Monitoring Network*, IUCN, Gland, Switzerland.

Johannes, R. E. (2002). The renaissance of community-based marine resource management in Oceania. *Annual Review of Ecology, Evolution, and Systematics, 33*, 317–357. https://doi.org/10.1146/annurev.ecolsys.33.010802.150524

Kittinger, J. N., Bambico, T. M., Minton, D., Miller, A., Mejia, M., Kalei, N., Wong, B., & Glazier, E. W. (2016). Restoring ecosystems, restoring community: Socioeconomic and cultural dimensions of a community-based coral reef restoration project. *Regional Environmental Change, 16*(2), 301–313.

Lebrec, M., Stefanski, S., Gates, R., Acar, S., Golbuu, Y., Claudel-Rusin, A., Kurihara, H., Rehdanz, K., Paugam-Baudoin, D., Tsunoda, T., & Swarzenski, P. W. (2019). Ocean acidification impacts in select Pacific Basin coral reef ecosystems. *Regional Studies in Marine Science, 28*, 100584. https://doi.org/10.1016/J.RSMA.2019.100584

Lemos, M. C., & Agrawal, A. (2006). Environmental governance. *Annual Review of Environment and Resources, 31*(1), 297–325. https://doi.org/10.1146/annurev.energy.31.042605.135621

Léopold, M., Thébaud, O., & Charles, A. (2019). The dynamics of institutional innovation: Crafting co-management in small-scale fisheries through action research. *Journal of Environmental Management, 237*, 187–199.

Liss, P. K. & Knight, F. W. (Eds.). (1991). *Atlantic Port Cities: Economy, Culture, and Society in the Atlantic World, 1650–1850.* University of Tennessee Press.

Lockwood, M., Davidson, J., Hockings, M., Haward, M., & Kriwoken, L. (2012). Marine biodiversity conservation governance and management: Regime requirements for global environmental change. *Ocean & Coastal Management, 69*, 160–172.

Lowe, P. K., Bruno, J. F., Selig, E. R., & Spencer, M. (2011). Empirical models of transitions between coral reef states: Effects of region, protection, and environmental change. *PLoS ONE, 6*(11). https://doi.org/10.1371/journal.pone.0026339

Magel, J. M. T., Burns, J. H. R., Gates, R. D. et al. (2019). Effects of bleaching-associated mass coral mortality on reef structural complexity across a gradient of local disturbance. *Scientific Reports, 9*, 2512. https://doi.org/10.1038/s41598-018-37713-1

Manzello, D. P. (2015). Rapid recent warming of coral reefs in the Florida keys. *Scientific Reports, 5*, 1–10. https://doi.org/10.1038/srep16762

Manzello, D. P., Berkelmans, R., & Hendee, J. C. (2007). Coral bleaching indices and thresholds for the Florida Reef tract, Bahamas, and St. Croix, U.S. Virgin Islands. *Marine Pollution Bulletin, 54*(12), 1923–1931. https://doi.org/10.1016/j.marpolbul.2007.08.009

McClanahan, T. R., Marnane, M. J., Cinner, J. E., & Kiene, W. E. (2006). A comparison of marine protected areas and alternative approaches to coral-reef management. *Current Biology, 16*(14), 1408–1413. https://doi.org/10.1016/j.cub.2006.05.062

McCook, L. J., Ayling, T., Cappo, M., Choat, J. H., Evans, R. D., De Freitas, D. M., Heupel, M., Hughes, T. P., Jones, G. P., Mapstone, B., & Marsh, H. (2010). Adaptive management of the Great Barrier Reef: A globally significant demonstration of the benefits of networks of marine reserves. *Proceedings of the National Academy of Sciences, 107*(43), 18278–18285.

Moberg, F., & Folke, C. (1999). Ecological goods and services of coral reef ecosystems. *Ecological Economics, 29*, 215–233. https://doi.org/10.1016/S0921-8009(99)00009-9

Morrison, T. H. (2017). Evolving polycentric governance of the Great Barrier Reef. *Proceedings of the National Academy of Sciences, 114*(15), E3013–E3021.

Morrison, T. H., Adger, N., Barnett, J., Brown, K., Possingham, H., & Hughes, T. (2020). Advancing coral reef governance into the anthropocene. *One Earth, 2*(1), 64–74. https://doi.org/10.1016/j.oneear.2019.12.014

NASA. (2020). *Global climate change: Vital signs of the planet* (H. Shaftel, R. Jackson, S. Callery, & D. Bailey, Eds.). Retrieved November 8, 2020, from https://climate.nasa.gov/

National Oceanic and Atmospheric Administration. (n.d.). *What is ocean acidification?* NOAA PMEL Carbon Program. https://www.pmel.noaa.gov/co2/story/What+is+Ocean+Acidification%3F

Ostrom, E. (1990). *Governing the commons: The evolution of institutions for collective action.* Cambridge University Press.

Ostrom, E. (2009). A general framework for analyzing sustainability of social-ecological systems. *Science, 325*(5939), 419–422. https://doi.org/10.1126/science.1172133

Ostrom, E. (2011). Background on the institutional analysis and development framework: Ostrom: Institutional analysis and development framework. *Policy Studies Journal, 39*(1), 7–27.

Pendleton, L., Comte, A., Langdon, C., Ekstrom, J. A., Cooley, S. R., Suatoni, L., Beck, M. W., Brander, L. M., Burke, L., Cinner, J. E., & Doherty, C. (2016). Coral reefs and people in a high-CO2 world: Where can science make a difference to people?. *PLoS ONE, 11*(11), e0164699.

Pendleton, L., Hoegh-Guldberg, O., Albright, R., Kaup, A., Marshall, P., Marshall, N., Fletcher, S., Haraldsson, G., & Hansson, L. (2019). The Great Barrier Reef: Vulnerabilities and solutions in the face of ocean acidification. In *Regional Studies in Marine Science* (Vol. 31, p. 100729). Elsevier B.V. https://doi.org/10.1016/j.rsma.2019.100729

Pietri, D., Christie, P., Pollnac, R. B., Diaz, R. and Sabonsolin, A. (2009). Information diffusion in two marine protected area networks in the central Visayas region, Philippines. *Coastal Management, 37*(3–4), 331–348.

Pomeroy, R. S. (1995). Community-based and co-management institutions for sustainable coastal fisheries management in Southeast Asia. *Ocean & Coastal Management, 27*(3), 143–162. https://doi.org/10.1016/0964-5691(95)00042-9

Pomeroy, R. S., Rebecca Rivera-Guieb, and Centre de recherches pour le développement international (Canada). (2009). *Fishery co-management a practical handbook.* CABI Pub.: In Association with the International Development Research Centre. Retrieved September 30, 2020, from http://site.ebrary.com/id/10119722

Precht, W. F., & Miller, S. L. (2007). Ecological shifts along the Florida reef tract: The past as a key to the future. In R. B. Aronson (Ed.), *Geological approaches to coral reef ecology* (pp. 237–312). Springer. https://doi.org/10.1007/978-0-387-33537-7_9

Rijke, J., Brown, R., Zevenbergen, C., Ashley, R., Farrelly, M., Morison, P., & van Herk, S. (2012). Fit-for-purpose governance: A framework to make adaptive governance operational. *Environmental Science & Policy, 22*, 73–84.

Roff, G. (2019). Reef accretion and coral growth rates are decoupled in Holocene reef frameworks. *Marine Geology, 419*, 106065. https://doi.org/10.1016/j.margeo.2019.106065

Rogers, A., Harborne, A. R., Brown, C. J., Bozec, Y. M., Castro, C., Chollett, I., Hock, K., Knowland, C. A., Marshell, A., Ortiz, J. C., & Razak, T. (2015). Anticipative management for coral reef ecosystem services in the 21st century. *Global Change Biology*, *21*(2), 504–514.

Schultz, L., Folke, C., Österblom, H., & Olsson, P. (2015). Adaptive governance, ecosystem management, and natural capital. *Proceedings of the National Academy of Sciences*, *112*(24), 7369–7374.

Scouter, D., Planes, S., Wicquart, J., Logan, M., Obura, D., & Staub, F. (2021a). Executive Summary. In *Status of coral reefs of the world: 2020*. Global Coral Reef Monitoring Network (GCRMN), International Coral Reef Initiative (ICRI).

Scouter, D., Planes, S., Wicquart, J., Logan, M., Obura, D., & Staub, F. (2021b). Chapter 8. Status and trends of coral reefs of the Australia region. In *Status of coral reefs of the world: 2020*. Global Coral Reef Monitoring Network (GCRMN), International Coral Reef Initiative (ICRI).

Scouter, D., Planes, S., Wicquart, J., Logan, M., Obura, D., & Staub, F. (2021c). Chapter 7. Status and trends of coral reefs of the East Asian Seas region. In *Status of coral reefs of the world: 2020*. Global Coral Reef Monitoring Network (GCRMN), International Coral Reef Initiative (ICRI).

Scouter, D., Planes, S., Wicquart, J., Logan, M., Obura, D., & Staub, F. (2021d). Chapter 9. Status and trends of coral reefs of the Pacific region. In *Status of coral reefs of the world: 2020*. Global Coral Reef Monitoring Network (GCRMN), International Coral Reef Initiative (ICRI).

Scouter, D., Planes, S., Wicquart, J., Logan, M., Obura, D., & Staub, F. (2021e). Chapter 3. Status and trends of coral reefs of the Red Sea and Gulf of Aden. In *Status of coral reefs of the world: 2020*. Global Coral Reef Monitoring Network (GCRMN), International Coral Reef Initiative (ICRI).

Smith, J. E., Brainard, R., Carter, A., Grillo, S., Edwards, C., Harris, J., Lewis, L., Obura, D., Rohwer, F., Sala, E., Vroom, P. S., & Sandin, S. (2016). Re-evaluating the health of coral reef communities: Baselines and evidence for human impacts across the central pacific. *Proceedings of the Royal Society B: Biological Sciences*, *283*(1822). https://doi.org/10.1098/rspb.2015.1985

Stanley Jr, G. D. (1981). Early history of scleractinian corals and its geological consequences. *Geology*, *9*(11), 507–511.

Sternlieb, F., Bixler, R. P., & Huber-Stearns, H. (2013). A question of fit: Reflections on boundaries, organizations and social–ecological systems. *Journal of Environmental Management*, *130*, 117–125.

Storlazzi, C. D., Reguero, B. G., Cole, A. D., Lowe, E., Shope, J. B., Gibbs, A. E., Nickel, B. A., McCall, R. T., van Dongeren, A. R., & Beck, M. W. (2019). Rigorously valuing the role of U.S. Coral reefs in coastal hazard risk reduction: U.S. Geological Survey Open-File Report 2019–1027, 42 p. https://doi.org/10.3133/ofr20191027.

Swierts, T., & Vermeij, M. J. A. (2016). Competitive interactions between corals and turf algae depend on coral colony form. *PeerJ*, *2016*(5). https://doi.org/10.7717/PEERJ.1984/SUPP-1

Veron, J. E. N., Devantier, L. M., Turak, E., Green, A. L., Kininmonth, S., Stafford-Smith, M., & Peterson, N. (2009). Delineating the coral triangle. *Galaxea, Journal of Coral Reef Studies*, *11*(2), 91–100. https://doi.org/10.3755/galaxea.11.91

Voosen, P. (2019). Scientists track Florida's Vanishing Barrier Reef. *Science*. https://www.science.org/content/article/scientists-track-florida-s-vanishing-barrier-reef#:~:text=%22We've%20lost%2090%25,wide%20branches%2C%20have%20nearly%20vanished

Wagner, D. E., Kramer, P., & Woesik, R. Van. (2010). Species composition, habitat, and water quality influence coral bleaching in southern Florida. *Marine Ecology Progress Series, 408*, 65–78. https://doi.org/10.3354/meps08584

Walker, B., Holling, C. S., Carpenter, S. R., & Kinzig, A. (2004). Resilience, adaptability and transformability in social–ecological systems. *Ecology and Society, 9*(2).

Walton, C. J., Hayes, N. K., & Gilliam, D. S. (2018). Impacts of a regional, multi-year, multi-species coral disease outbreak in Southeast Florida. *Frontiers in Marine Science, 5*, 323.

Weis, V. M. (2008). Cellular mechanisms of Cnidarian bleaching: Stress causes the collapse of symbiosis. *Journal of Experimental Biology, 211*(19), 3059–3066.

Wells, S., Ray, G. C., Gjerde, K. M., White, A. T., Muthiga, N., Bezaury Creel, J. E., Causey, B. D., McCormick-Ray, J., Salm, R., Gubbay, S., & Kelleher, G. (2016). Building the future of MPAs – Lessons from history. *Aquatic Conservation: Marine and Freshwater Ecosystems, 26*(S2), 101–125.

Westcott, D. A., Fletcher, C. S., Kroon, F. J., Babcock, R. C., Plagányi, E. E., Pratchett, M. S., & Bonin, M. C. (2020). Relative efficacy of three approaches to mitigate Crown-of-Thorns Starfish outbreaks on Australia's Great Barrier Reef. *Scientific Reports, 10*(1), 1–12. https://doi.org/10.1038/s41598-020-69466-1

White, A. T., & Vogt, H. P. (2000). Philippine coral reefs under threat: Lessons learned after 25 years of community-based reef conservation. *Marine Pollution Bulletin, 40*(6), 537–550. https://doi.org/10.1016/S0025-326X(99)00243-X

Wilson, B., Fan, C. M., & Edmunds, P. J. (2020). An unusual microbiome characterises a spatially-aggressive crustose alga rapidly overgrowing shallow Caribbean reefs. *Scientific Reports, 10*(1), 1–13. https://doi.org/10.1038/s41598-020-76204-0

Woodhead, A. J., Hicks, C. C., Norström, A. V., Williams, G. J., & Graham, N. A. (2019). Coral reef ecosystem services in the Anthropocene. *Functional Ecology, 33*(6), 1023–1034.

Young, O. R. (2008). Building regimes for socioecological systems: Institutional diagnostics. In O.R. Young, L. A. King, & H. Schroeder (Eds.), Institutions and environmental change:Principal findings, applications, and research frontiers (pp. 115–144). Cambridge, MA: MIT Press.

Part 2

CASE STUDIES

Chapter 5

THE CASE OF THE FLORIDA REEF TRACT: BUREAUCRACIES, PARTICIPATION AND MANAGING NOVEL ECOSYSTEMS

By Kasen Wally and Kelly Dunning

The concept of a pristine coral reef, untouched and unspoiled, feels like a thing of the past. While we know they must exist and hear stories about their majestic beauty, they feel far away. Depending upon how you define a healthy coral reef, it is debatable if the word "pristine" accurately describes coral reefs at all. While beautiful and intact reef ecosystems certainly exist, are these reefs truly unspoiled? They may encounter the odd abandoned piece of fishing gear, the curious lack of a once abundant fish species or unseasonably warm ocean temperatures. Thus, are these places still pristine, or have they shifted to become something new?

Many people might not know that until fairly recently, they could find a healthy tropical coral reef at the southern tip of Florida in the United States. Until the 1980s, this 360 linear mile barrier reef—the Florida Reef Tract—was world-renowned as a premier dive site, showcasing rich collections of species (known as assemblages) and dense coral cover. In fact, the Florida Reef Tract is the third-largest barrier reef in the world.

Growing up in South Florida in the 1990s, stands of elkhorn coral on the reefs of the Florida Keys National Marine Sanctuary (FKNMS) were expected to be seen on dives, with their yellow, branching arms stretching through the clear water upward toward the light. In summer of 2022, while on a research dive, I recall crying out and pointing to a stand of elkhorn to a student simply because it has become such a rare sight in this short time, that it warrants a celebration.

Over the last 40–50 years, close to 90 percent of the living coral cover that comprised the Florida Reef Tract has been lost (Figure 5.1), and new species assemblages have begun to take their place (National Oceanic and

Figure 5.1 A time series of coral reef decline over recent decades at Grecian Rocks Reef, Key Largo, Florida. Source: Saint Petersburg Coastal and Marine Science Center, 2017.

Atmospheric Administration, 2022; Toth et al., 2019). Due to these significant changes, new forms of management have emerged to try and conserve what coral remains.

In this chapter, we explore two of the most visible and important case studies of coral reef management institutions from southern Florida: the case of the FKNMS and the case of the Southeast Florida Coral Reef Initiative (SEFCRI). These two management organizations have arisen under different circumstances along the Florida Reef Tract, emerging as strikingly different institutions next to some of the largest, fastest growing cities in the United States, and thus, they are visited often and by many. By comparing them, we are able to generate valuable insights into how decision-makers can approach management, form linkages and cooperate toward shared goals at a regional scale. The public and other stakeholders are frequently invited to play a role in the way these institutions manage coral reefs, although the participatory nature of these institutions couldn't be more different.

The Ecological Context

Much of Florida's economy depends on the Florida Reef Tract, which generates billions of dollars each year via recreational and commercial fishing, the diving and tourism industry and other businesses in South Florida (Wynveen et al., 2013). Due to this relationship, many diverse stakeholders are significantly invested in the health and productivity of this vital ecosystem (Cinner et al., 2009). Despite their importance, coral reefs are ecologically fragile and face many anthropogenic threats such as climate change, water pollution, overfishing and mechanical damage (e.g., ship groundings or anchor damage) (Frys et al., 2020).

The reefs of South Florida are very popular, but it is their popularity that has resulted in their degradation over time. There are numerous environmental stressors, such as highly developed coastlines with dune ridges covered in hotels and condos—development that enables one of the fastest growing populations in the United States. These large numbers of people moving to Florida continually set visitation records for the reef, engaging in boating, recreational fishing and diving. In other words, large numbers of people move to or vacation to Florida to experience nature which is being degraded through its use by these same people. White sandy beaches would not be as dazzling as they are without the reef system. World-class fishing would not exist without the habitat provided by the Florida Reef Tract which shelters fish at all life stages. Thus, a robust economy and coastal beauty attract residents, while creating environmental damage due to the sheer numbers who continue to move to Florida. These stressors can often compound, increasing the rate of degradation and making conservation difficult for managers (Dustan, 2003). For example, coral disease, a result of poor water quality and other factors, can kill coral and reduce the structural complexity of the reef; in turn, this impairs the ability of the reef to act as a buffer from massive waves during Florida's hurricanes, which can increase damage to coastal infrastructure, like beachfront housing.

While almost two-thirds of the Florida Reef Tract is adjacent to the Florida Keys, an island archipelago in Monroe County, the remaining third extends northward to Martin County, east of Lake Okeechobee (Figure 5.2) (Douglas, 2020). As one can see in Figure 5.2, the Reef Tract encompasses an enormous area, and many local factors, like population density, contribute to its overall health and stability. The reasons for this are many, but take for instance the fact that more people means more lawns, and likewise more lawn fertilizer (e.g. nitrogen and phosphorus). Fertilizers wash into coastal waters with rain. These nutrients wash out to the coral reef, increasing disturbances on an ecosystem that requires low nutrients. Nutrients may enable algal blooms in the water or on the coral itself, blocking coral's access to light which it requires to make part of its food. Factor in other human uses like fishing and diving, and the disturbances and that result, and it is no secret why reefs are struggling.

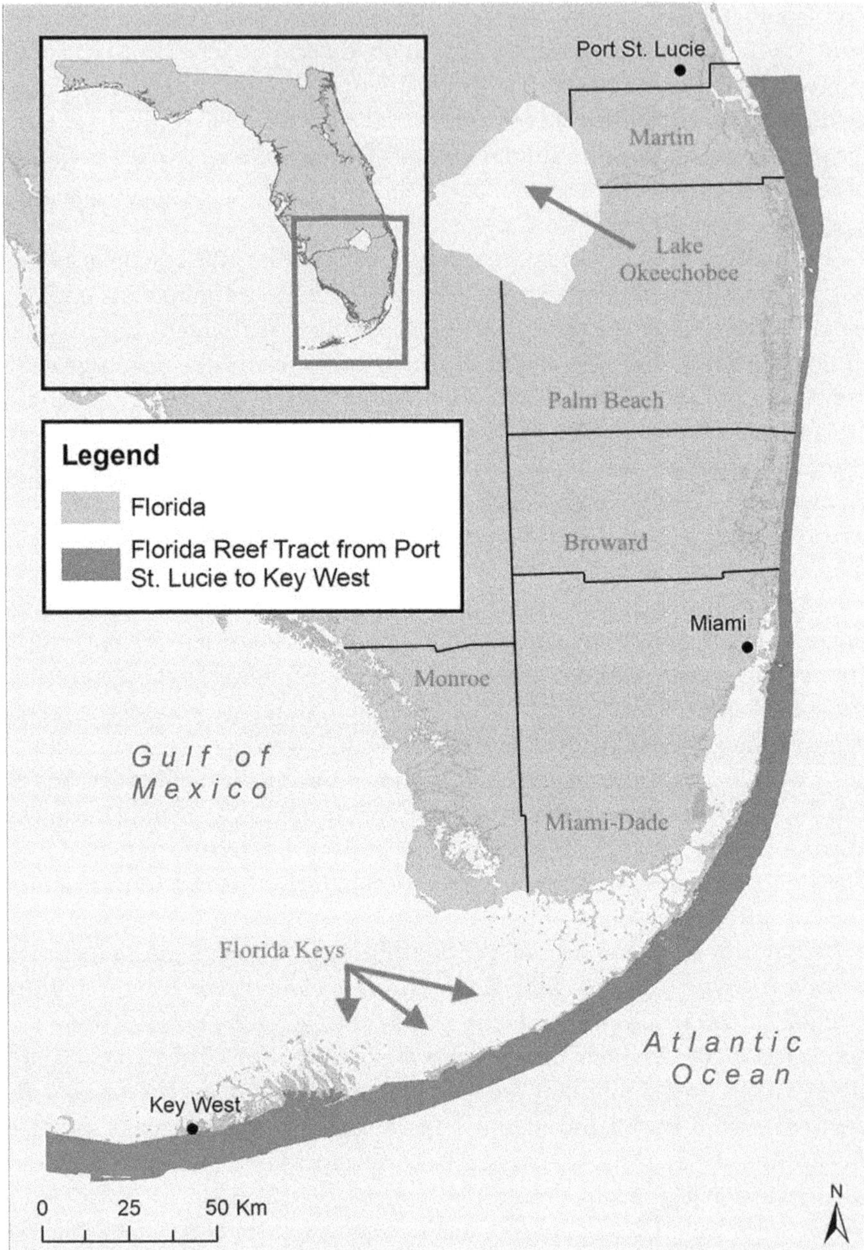

Figure 5.2 A depiction on the full extent of the Florida Reef Tract. Source: Yates et al., 2019. This image was originally sourced from the U.S. Geological Survey (Yates et al., 2019) and adapted to fit the context of this chapter.

Remarkably, until recently only the lower two-thirds of the Florida Reef Tract was managed under a comprehensive management plan, which was (and still is) administered by the FKNMS. In contrast, the remaining northern third of the Reef Tract lacked a management plan until 2003. At that time, the Florida Department of Environmental Protection, a state agency tasked with implementing laws that protect natural resources, helped form a governance organization to manage coral reefs known as the SEFCRI. SEFCRI began to manage this section of the reef cooperatively using a type of management plan called the Local Action Strategy (Reisewitz & Harper, 2013).

It may be surprising that in a well-resourced country like the United States that an ecosystem as precious as coral reefs would lack a management plan until fairly recently. In its absence, a patchwork of federal and state endangered species laws for threatened and endangered species of fish and coral were the strongest rules governing how people could use the resources of the reef. But rules are only as good as their enforcement. It would be impossible for federal or state game wardens to patrol every stretch of reef reminding people not to touch or kick the reef while diving or snorkeling, not to anchor on it, and not to feed or chase wildlife, but this type of behavior is surprisingly common, and reaches a crescendo during heavy tourist seasons such as Lobster Mini Season.[1] The governance of Florida reefs is continually evolving to better adapt to widespread human use and environmental stressors.

One of the earliest bleaching events in the late 1980s compelled some of the management changes that we recognize today. In 1987, several massive coral bleaching events significantly disrupted the reef ecosystem and destroyed habitat (Manzello, 2015). Coral bleaching is a stress response, usually seen with rising ocean temperatures (a consequence of climate change), that causes corals to expel symbiotic microorganisms they rely upon to survive. When this happens, the corals often die (Wagner et al., 2010). Unfortunately, the damage from these events was particularly severe among two species of branching coral, commonly known as Elkhorn (*Acropora palmata*) and Staghorn coral (*Acropora cervicornis*), which used to be the dominant coral species on the Florida Reef Tract (Toth et al., 2019). This damage was not caused by climate change and rising ocean temperatures alone. It occurred in conjunction with a decrease in fish and invertebrate species, and a disease known as White

---

1 A period in late July where recreational spiny lobster season opens for several days before the traditional season opens later in the year. Lobster mini season has become so popular that Fish and Wildlife runs public outreach campaigns encouraging public safety due to the number of diving and boating accidents.

Band Disease, which was the suspected result of regional water pollution in the 1970s (Jackson et al., 2014). Unfortunately, this compounding stress moved quickly; by 2010, coral cover on the Florida Reef Tract had dropped below 10 percent total coral cover, and by 2020 a staggering 98 percent loss (Chinn, 2020).

As the coral species which were once abundant on the Florida Reef Tract disappeared, they began to be replaced by opportunistic species like fleshy macroalgae, sponges and other species (Alvarez-Filip et al., 2013; Cramer et al., 2020; Loh et al., 2015). Refer back to Figure 5.1 and notice that the corals in the upper-left panel have been replaced by sea fans and soft corals in the bottom-right panel. While these species are often faster-growing and more stress-tolerant, they fail to provide many ecosystem services that reef-building corals do (Burman et al., 2012; Kuffner & Toth, 2016; Storlazzi et al., 2019).

Another significant change in Figure 5.1 is the overall flattening of the reef. As previously mentioned, corals provide significant structural complexity. Without three-dimensional structural complexity, coral reefs lose much of the functionality that enables the suite of ecosystem services they provide, such as reducing the size of storm surge during a hurricane or the erosion capabilities of the tides (Burman et al., 2012; Kuffner & Toth, 2016; Storlazzi et al., 2019). As these reef-building species disappear and are replaced, biological, chemical and physical processes of reef erosion and deterioration accelerate, resulting in a net loss of calcium carbonate and the eventual overall flattening of the reef (Perry & Alvarez-Filip, 2019; Perry et al., 2015; Toth et al., 2019). This adversely affects both humans (e.g., through coastal erosion and storm damage) and reef species like fish and invertebrates (e.g., through decreased protection from predation) (Graham & Nash, 2013).

In addition to loss of coral, in the 1980s the long-spined sea urchin (*Diadema antillarum*) was nearly eradicated by an unknown pathogen (Cramer et al., 2020; Weil et al., 2006). These urchins are keystone species, meaning that they are essential to the ecosystem's functioning (Nuñez & Dimarco, 2010). In conjunction with herbivorous (or plant and algae eating) fish, long-spined sea urchins are responsible for helping to control fleshy macroalgae populations, which can easily out-compete the slower-growing corals, starving them and further damaging the ecosystem (Lessios, 2016; McManus & Polsenberg, 2004). Today, the urchin population has yet to fully rebound. As of 2016, the population had reached approximately 12 percent of their historic numbers, indicative of the severity of the die-off and the lasting impact on the ecosystem (Lessios, 2016). Furthermore, their natural recovery has been significantly hampered by the reduced structural complexity seen on the reef, partially a consequence of their disappearance (Bodmer et al., 2021; Pilnick et al., 2021).

While these urchins are making a comeback, their disappearance created "free real estate" for encroaching organisms (Jackson et al., 2014).

Whether herbivorous fish species were overfished in Florida when the sea urchin population began to decline remains the topic of some debate (Jackson et al., 2014; McClenachan, 2013). If so, reductions in fish populations may have contributed to fleshy macroalgae abundance in the region following the disappearance of the sea urchins. Regardless, in addition to the disappearance of keystone herbivorous species on the Florida Reef Tract, overfishing has become a significant problem in Florida. The removal of top predators, including groupers, snappers and sharks, has likely caused one or more trophic cascades, affecting reef-fish species assemblages, including herbivorous fish (McClenachan, 2009; Pandolfi et al., 2003; Precht & Miller, 2007). As a result, these changes have indirectly contributed to alterations to invertebrate species compositions, such as increasing populations of encroaching sponges, which would otherwise become prey for herbivorous species (Loh et al., 2015).

The Florida Reef Tract we see today is not the thriving ecosystem it once was. It has now entered a "novel" state of existence which can be defined by its many alterations to species configurations, ecological interactions and overall functionality (Clement & Standish, 2018; Hobbs et al., 2009; Toth et al., 2019). Therefore, we must consider how management has evolved to tackle these problems through the creation of organizations like FKNMS and SEFCRI; and how management must change in the decades ahead. With evolving management, considerations of how stakeholders are invited as co-managers provide insights on best management practices for global reefs facing similar environmental changes.

Background

Governance of the FKNMS

The FKNMS was established in 1990 under the Marine Protection, Research, and Sanctuaries Act of 1972. However, its creation was in the works for years. Initial discussions on the conservation of the Florida Reef Tract began as early as 1957 and resulted in the formation of John Pennekamp Coral Reef State Park, the first underwater park of its kind in the United States (U.S. Department of Commerce, 2007).

Following many of the changes to the Florida Reef Tract that we have discussed so far, including coral bleaching, widespread disease and declines among keystone species, people began to take notice of the Reef Tract's deteriorating condition, which sparked environmental reforms throughout the 1970s and 1980s (Frys et al., 2020; Gregg, 2013). In 1975 the Key Largo

National Marine Sanctuary was established, and the Looe Key National Marine Sanctuary followed shortly after in 1981 (National Oceanic and Atmospheric Administration, 2015). While these protected areas were incremental steps toward marine conservation, they also paved the way for the eventual creation of the FKNMS that we see today.

By 1988, as oil and gas prospectors fought to secure leases for oil drilling on the Florida Reef Tract, a series of ship groundings damaged sensitive habitat and ultimately led to feasibility to assess whether the existing protected areas from the 1980s should be expanded (U.S. Department of Commerce, 2007). These reports ultimately highlighted the precarious state of the Florida Reef Tract and generated bipartisan support in the U.S. Congress, resulting in the formation of the FKNMS on November 16, 1990 (U.S. Department of Commerce, 2007).

Today, the FKNMS is a nationally recognized management organization that is responsible for the stewardship of the southern portions of the Florida Reef Tract. The FKNMS is federally administered by NOAA and co-managed at the state level by the Florida Department of Environmental Protection (DEP). The sanctuary itself is enormous, including 2,800 square nautical miles. Currently, there are five types of marine zones within the sanctuary that support various types of resource uses and protections for habitat and wildlife. Sanctuary-wide regulations exist and are an important part of FKNMS management. Such regulations include limitations on human-wildlife interactions with endangered species (corals) and enact marine zoning with stricter regulations such as complete restrictions on fishing (e.g., no-take zones). These regulated spatial extents only account for a small fraction of the FKNMS's total area. (Office of National Marine Sanctuaries, 2019).[2]

The Sanctuary's catalyst, the FKNMS and Protection Act, also required a Comprehensive Management Plan, which was put into place in 1997 and revised in 2007 containing information on stressors and interventions to address stressors (Baker, 1999; Gregg, 2013). As of 2021, the Comprehensive Management Plan is undergoing a third revision based on the outcomes of another draft environmental impact statement known as the FKNMS Restoration Blueprint.[3] As stipulated by Congress, the current iteration of the Comprehensive Management Plan aims to reduce environmental degradation in critical areas by reducing exploitation, protecting vulnerable species

2 A complete overview and interactive map of the Sanctuary's zones can be found here: *"Exploring Sanctuary Zones and Regulations."* https://arcg.is/0qGSe0

3 More information about the Restoration Blueprint can be found here: *"Florida Keys National Marine Sanctuary Restoration Blueprint."* https://floridakeys.noaa.gov/blueprint/

and minimizing conflicts among sanctuary visitors through an evolving process of marine zoning (Office of National Marine Sanctuaries, 2019). The ongoing public engagement process involved in the Restoration Blueprint is the result of years of robust public engagement, education, public comment periods and conversations between core stakeholder groups and managers. Managers are considering cutting-edge ecosystem and climate science to make management decisions grounded in the essential needs of stakeholders such as diving companies, commercial and recreational fishers.

The management plan is continually monitored and reviewed through federal and state agencies in conjunction with the Sanctuary Advisory Council. The Sanctuary Advisory Council is made up of representatives from the community (such as teachers, commercial and recreational anglers, business owners and environmental non-profits); government personnel such as representatives from the Monroe County government; universities, and other organizations with vested interests in the health of the Florida Reef Tract (Sleasman, 2009). Based on the input of the Sanctuary Advisory Council, the regulations set forth by the FKNMS are strictly enforced by the Florida Fish and Wildlife Conservation Commission in partnership with the NOAA Office of Law Enforcement and the U.S. Coast Guard (U.S. Department of Commerce, 2007).

The Sanctuary Advisory Council is a unique organization to ensure participatory management. While skeptics note that its decisions, made through members casting votes, are *advisory* only, there are many serving that have lived next to the FKNMS for decades. Many are pillars of the nearby community, including a seat meant as a stand-in for a general member of the public ("member of the public at large"), held by a notable local science teacher. The Council is a novel way to ensure federal bureaucratic decision-makers hear all voices in their decision-making process.

Initially, the FKNMS Comprehensive Management Plan was non-regulatory, focusing instead on public education, marking channels to avoid boat collisions with the reefs, ensuring environmental surveying to designate important reef habitat, and the installation of mooring buoys to prevent anchor damage. However, through time, it has developed to include (1) many new management areas, such as Sanctuary Preservation Areas (e.g., SPAS) meant to reduce fishing and taking of living marine life, and (2) regulations that help to manage water quality and social-ecological interactions (Gershman et al., 2012; Office of National Marine Sanctuaries, 2019). For example, since its implementation in 1997, NOAA has expanded the sanctuary to include the Dry Tortugas Ecological Reserve, added no-discharge regulations in state and federal waters to improve water quality, and implemented numerous monitoring programs to track coral bleaching, coral disease and other indicators of ecosystem health (Office of National Marine Sanctuaries, 2019).

Governance of the SEFCRI

Thirteen years after the creation of the FKNMS, the more northerly reefs from Biscayne National Park in the South to Port St. Lucie in the North were placed in the SEFCRI MPA, formed in 2003. It now includes over 70 organizations that work as partners and has implemented over 140 individual conservation projects through their Local Action Strategies, or conservation interventions, which identify opportunities to improve the health of these northern reaches of the Florida Reef Tract. Projects include water quality sampling, invertebrate species surveys and socioeconomic studies that help inform decision-making (Florida Department of Environmental Protection, 2004; Gregg, 2013).

SEFCRI was created through the United States Coral Reef Task Force (or "Task Force"), established in 1998 to protect not only biodiversity, but also health, cultural heritage and socioeconomic values of U.S. reefs (Florida Department of Environmental Protection, 2004). The Task Force includes seven representatives of both U.S. territories and states with coral reef ecosystems. The Task Force is responsible for creating conservation planning initiatives, called Local Action Strategies, for each of its members in places like Hawaii, American Samoa and Puerto Rico. The Local Action Strategies essentially acts as a management plan akin to FKNMS's management plan.

An important difference is that the FKNMS Management Plan draws on legislative authority and regulatory enforcement, while the Local Action Strategy of SEFCRI proposes a constellation of projects which can be adopted by federal, state, local or non-governmental partners with no sole agency or law in charge. Ultimately, SEFCRI fills a gap between formal governance like that found within the FKNMS with non-governmental partners like universities and non-profits. Therefore, SEFCRI's network includes many different government agencies, non-governmental organizations, universities and private partners with one thing in common—a mission to protect the Florida Reef Tract while maximizing transparency and involvement of interested stakeholders (Southeast Florida Coral Reef Initiative, 2019).

SEFCRI is managed by the Florida Department of Environmental Protection under their Coral Reef Conservation Program. The SEFCRI MPA itself covers a significant range, including Miami-Dade, Broward, Palm Beach and Martin Counties. The SEFCRI team operates using focus groups that tackle the five primary subjects within their Local Action Strategy. These focus areas include: (1) Awareness and Appreciation, (2) Fishing, Diving and Other Uses, (3) Land-Based Sources of Pollution, (4) Maritime Industry and Coastal Construction Impacts and (5) Reef Resilience (Kerrigan, 2017; Reisewitz & Harper, 2013). Each focus group is led by representatives, termed "Navigators," who are responsible for setting goals, objectives, actions, management priorities, budget development, building implementation teams and tracking progress (Florida Department of Environmental Protection, 2004).

Navigators may be agency-affiliated (where they represent their agency within SEFCRI), or they may also be non-agency Navigators who represent local stakeholder views and exist as primary points of contact for their constituents. In this way, SEFCRI creates a network of participants, all working toward common goals (Florida Department of Environmental Protection, 2004). Like the FKNMS, the Florida Department of Environmental Protection's management decisions rely on a transparent process that encourages stakeholder inclusion. An advisory board (known as the Technical Advisory Committee) made up of the leading research scientists in various fields related to reef management assists the SEFCRI team and the Florida Department of Environmental Protection in identifying and implementing priority actions to reduce key threats to coral reef resources in the SEFCRI MPA (Florida Department of Environmental Protection, n.d.).

To select priority actions within each of the five focus areas, SEFCRI has developed guidelines that assess any project's scope, feasibility, relevance, requirements and ethics (Florida Department of Environmental Protection, 2004). The prioritization process helps ensure that any projects undertaken in new iterations of the Local Action Strategy are the best-fit for the issues at hand. For example, in 2017, SEFCRI enacted a project known as the Reef Resilience project, which produced critically important research on reef health co-funded by important partners such as NOAA's Coral Reef Conservation Program, the Florida Department of Environmental Protection and The Nature Conservancy's Florida office. This research showed that the northern Florida Reef Tract was extremely vulnerable to the effects of climate change and suggested management actions that could address critical risk factors (Maynard et al., 2017). Typically, each of the proposed management actions aims to be completed within a three-year time frame. The current iteration of the Local Action Strategy, also released in 2017, is still underway, meaning there can be some delays (Florida Department of Environmental Protection, 2004; Kerrigan, 2017). There is no proposed end to the SEFCRI mission. Presumably, operations will continue long-term, given that funding is appropriated and the ecological situation necessitates further action.

Our Approach

This chapter provides the first formal analysis of the governance of these two critically important MPAs which serve as models for coral reef conservation in the United States. To perform this analysis, we use the concept of *institutions* to describe the rules that govern the behavior of actors enacting coral reef conservation. Actors are the governmental and non-governmental participants who practice environmental governance (Pahl-Wostl, 2015). In this chapter, when we describe the FKNMS and SEFCRI as institutions, we mean the collection of actors (e.g., governmental agencies, non-governmental organizations and public stakeholders) who rely on different types of institutions to cooperate and function independently.

There are primarily two types of institutions. First, *formal institutions* rely on rules and laws that are legally enforced, typically by state actors (i.e., governmental organizations) (Pahl-Wostl, 2009). Second, *informal institutions* rely on public discussions and cultural norms that are not typically legally enforced. These types of informal arrangements typically include non-state actors (i.e., private companies, non-profits and individuals) (Pahl-Wostl, 2009). While these actors may play significant roles in decision-making, they typically lack the regulatory authority found in formal institutions (Carlisle & Gruby, 2019). With this information in mind, we incorporate the concepts of institutions to help explain how MPAs vary across spatial extents in the Florida context using the FKNMS and SEFCRI. We provide comparisons of the strategies used to manage the Florida Reef Tract across the two different MPAs to help understand the varying coral reef management systems used in the United States and how these systems can meet the future challenges of climate change.

Given the changes that we have seen on the Florida Reef Tract, it is unlikely that it will ever fully recover to its historical state (Graham et al., 2013). Additionally, conventional management, like the implementation of MPAs, is not likely to curb the effects of climate change (Graham et al., 2014; Hughes et al., 2017b). To meet the needs of this changing ecosystem, managers must implement new strategies and tactics to improve conservation (Braverman, 2020; Graham et al., 2013).

One way managers can improve decision-making is through adaptive governance, which relies on iterative learning, or simply put, learning as we go, to improve the management of socio-ecological systems over time (Folke et al., 2005). There are many different examples of adaptive governance incorporated into coral reef management, such as exploring trade-offs between coastal infrastructure and ecological concerns (Dunning, 2021). Meaning, decision-makers see what happens when they allow rapid development of hotels and tourism infrastructure along coral reef coastlines. If the water quality becomes too poor, and reefs begin to suffer mortality events, new interventions can be put in place like slowing the speed of development or requiring construction companies to stop sediment inputs into coastal waters. Our research in this chapter is among the first to ask how an adaptive governance approach can fully consider the changes, shifting baselines and contemporary status quo of the Florida Reef Tract?

The novel ecosystem framework

To help us explore the differences between our two cases, and the challenges of adaptive governance, we use the concept of *novel ecosystems*, which are defined by dramatic changes to species compositions and alterations to the fundamental processes which support ecosystem functions (Clement & Standish, 2018; Graham et al., 2014; Hobbs et al., 2013). The Florida Reef Tract fits this definition perfectly, where changes to ecosystems include shifts in species distribution and composition, negatively affecting ecosystem functionality.

There are five essential steps for managing novel ecosystems. These are areas that, if ignored by managers, would lessen the effectiveness of strategies and interventions. We operationalized these concepts to explore differences between the FKNMS and SEFCRI. These concepts were: (1) *Administrative Competence*, (2) *Buffering*, (3) *Culture and Norms*, (4) *Framing* (which we termed "*Issue Framing*") and (5) *Power and Authority* (Clement & Standish, 2018).

In the context of coral reef management and novel ecosystems, the concept of *Administrative Competence* is ubiquitous (Anthony et al., 2020; Clement et al., 2016; Morrison et al., 2020; Rijke et al., 2012). Administrative competence means that management actions effectively conserve the ecosystem. Examples include: implementing new MPAs; improving protocols for data collection; or ensuring that the right people are appointed to the correct positions. Without a doubt, agencies, organizations and managers' ability to handle complex, unpredictable and open-ended problems (like the management of the Florida Reef Tract) often call for extensive management capabilities and practical strategies (Head & Alford, 2015).

Next, Buffering asks if management is adequately prepared for the future. Managing uncertainty is an integral part of managing novel ecosystems because unpredictability is inherent to complex socio-ecological systems (Allen & Gunderson, 2011; Hobbs et al., 2013). How much will the earth's climate warm in 50 years? In 150 years? How will reef species respond? Managers can do their best to predict the future, but they must be prepared for uncertain outcomes. Examples of buffering include discussing contingency plans for possible climate scenarios, expanding protections to habitats closely linked to and essential for the health of the reef tract (like seagrass habitats) or implementing long-term monitoring programs to lessen uncertainty. Ultimately, including the concept of buffering in management helps prepare for the future (Clement & Standish, 2018).

The concept of *Culture and Norms* asks whether management includes stakeholders, engages the public and encourages a shared understanding of novel ecosystems. The cultural value of ecosystems must be considered in decision-making (Gee et al., 2017). Understanding how and why stakeholders value ecosystems help managers to incorporate attitudes, beliefs and behaviors into management strategies (Ives & Kendal, 2014). Examples include: considering access to an ecosystem as a form of cultural heritage in South Florida, creating public outreach campaigns to increase awareness or giving stakeholders a primary role in decision-making processes. Clement and Standish (2018) also emphasize that helping everyone be on the same page about what a novel ecosystem needs and how it differs from historical conditions can help managers meet objectives. South Florida life is often oriented toward the reef, whether spending time boating, fishing or diving on the weekends or enjoying the white sandy beaches that reefs help create, people in these communities have a way of life strongly linked to the reefs. Culture and Norms asks whether decision-makers effectively link this way of life to education and engagement.

Issue Framing considers whether management adequately incorporates man-made impacts and novel solutions into their strategies. Examples include managers discussing the implications of new species compositions (e.g., the loss of hard coral along the reef tract), mentioning a need for adaptive governance and acknowledging new baselines. Ultimately, framing coral reef degradation as only a biological problem rather than an anthropogenic problem tied to social and economic systems may fail to generate appropriate interventions (Morrison et al., 2020). Reframing perceptions and understandings of novel ecosystems enable more innovative management strategies to be implemented, incorporating cultural values and building adaptive capacity among institutions (Clement & Standish, 2018).

Finally, *Power and Authority* asks whether management is well organized, capable, respected and willing to consider change. This concept often deals with perceptions surrounding environmental governance (i.e., what people think about how management operates). Understanding power structures, policy-making and governance are essential for deciphering legislation, rules, responsibilities and roles (Clement & Standish, 2018). Examples of power and authority include managers discovering ways to increase inter-agency cooperation, managers working with the public to increase the perceived legitimacy of the organization or implementing new fisheries regulations. Public trust and the perceived legitimacy of management are essential parts of governance (Barrow & Murphree, 2001). Public support improves public compliance with rules, laws and regulations and should be assessed when considering novel ecosystem governance (Borrini-Feyerabend & Hill, 2015; Thaker et al., 2019; Young, 2016).

Using 1122 statements from news documents, public meeting transcripts and stakeholder interviews, we quantified instances of each concept and using those numbers, we were able to measure their relative importance to the FKNMS and SEFCRI. For instance, we found that Administrative Competence was not only the most important concept for managers in both organizations, but in equal measure. Additionally, we looked at criticisms of each concept with the organizations to determine pathways for improved management in the future.

What We Found

The most important difference we found between the two MPAs on the Florida Reef Tract was the degree to which they had a hierarchy. Foremost, when multiple actors (state, non-state, or both) interact, they form actor networks. These actor networks operate using "modes of governance" (Pahl-Wostl, 2009). Modes of governance include bureaucratic hierarchies (formal institutions of government, where state actors make decisions) and networks (state and non-state actors collaborate to make decisions). The FKNMS is primarily a top-down, bureaucratic hierarchy, whereas SEFCRI is mainly a network. This is not to say that these modes of governance are mutually exclusive, just that the FKNMS and SEFCRI approach management differently. One of our interview respondents tasked with reef management was able to explain this core difference:

> The FKNMS is managed nationally, so there is an inherent entity that is making decisions, whereas SEFCRI exists strictly because there are so many different management entities [...] the lines [become] blurred.

Remember, while SEFCRI is administered by a state agency, The Florida Department of Environmental Protection, the organization must still coordinate across four different counties, each with different socio-ecological concerns and priorities. For example, north of the sprawling cities of Miami and Fort Lauderdale, Martin County is home to more rural areas, public lands and agriculture. These types of land-use impact water quality in different ways than large population centers—namely via sources of pollutants, like agricultural runoff versus waste-water outflows. Additionally, SEFCRI relies on partner organizations within their actor-network to implement projects from the Local Action Strategy.

In contrast, the FKNMS is contained within Monroe County, and exists to implement state-mandated regulations legally enforced by clearly defined state actors in leadership roles, such as the Florida Fish and Wildlife Conservation Commission and the United States Coast Guard.

Direct actions taken on the ecosystem (and the results of those actions) become apparent in viewing examples of how managers tackle conservation goals. For instance, within the FKNMS, 27 wildlife management areas have been designated to protect sensitive wildlife habitats such as bird and turtle-nesting areas. Human access is heavily regulated in these areas, and restrictions are legally enforced (Gershman et al., 2012).

In contrast, SEFCRI has focused on data collection, which obviously lacks much of the same regulatory authority, but has generated some profound results. For example, one of SEFCRI's initial projects established the Southeast Florida Coral Reef Evaluation and Monitoring Project, which provides detailed annual reports on the status and trends of the region (Florida Department of Environmental Protection, 2022). Differences like these will become clearer as we walk through how SEFCRI and the FKNMS incorporated each of the novel ecosystem concepts, beginning with Administrative Competence.

Administrative competence: Are we actually doing conservation?

For the FKNMS, Administrative Competence, or managers achieving effective conservation, is enacted through their use of top-down leadership and the enforcement of regulations, such as managers making changes to the existing management plan, new proposals for protected areas (e.g., no-take areas), water quality protection programs and permitting for artificial reef structures. The implementation of these measures is often action-oriented, meaning that straightforward conservation actions are being taken. For example, when an experimental protected area (the Dry Tortugas Research Natural Area) was identified by the FKNMS as having been an effective MPA, it was quickly reinstated—an example of active resource management:

> These data, which were presented to Florida Fish and Wildlife Conservation Commissioners, show that the Dry Tortugas Research Natural Area was an effective MPA that now has been reinstated for another 20 years.—Research Assistant, Florida Fish and Wildlife Conservation Commission

Protected areas like these serve as concrete examples of management actions ongoing in the FKNMS, which reflect high levels of Administrative Competence. To further illustrate this, consider this quote from the Sanctuary Superintendent in 2016:

"We have some new science and condition reports to show what's working and where changes may need to be made." This quote was about expanding areas known as Sanctuary Protected Areas, and it demonstrates an iterative and adaptive process aimed at improving results.

On the other hand, while Administrative Competence was found to be equally important to SEFCRI, its approach differed in practice, focusing more on preventative measures. Managers seemed to prioritize data collection, such as curbing harmful nutrient flows from agricultural run-off and surveying for coral disease. An important development in 2014, the emergence of a new coral disease, Stony Coral Tissue Loss Disease, off the coast of Miami-Dade County and its rapid spread, became one of the most devastating causes of coral decline on the Florida Reef Tract today, and an important focus of preventative measures (Walton et al., 2018). Thus, it was often mentioned within our data and is an important ecological concern, especially within SEFCRI's understanding of the northern Reef Tract.

SEFCRI operationalized Administrative Competence by identifying voluntary or collaborative management actions (vs. top-down decisions from a government agency), opportunities for cooperation, new management goals, and the implementation of monitoring, education and outreach programs. SEFCRI also commonly focused on the feasibility of preventative measures to reduce stressors, like treatment options for coral disease, through processes of structured decision-making.[4] Decision-makers involved with SEFCRI's management frequently noted the need to transition management to a style with more in common with the FKNMS. A quote from SEFCRI's Technical Advisory Committee meeting minutes in 2015 highlights the extent of its focus on preventative measures and data collection while also highlighting a need to shift toward more active (regulatory) management practices:

> We have data and research that is sitting on a shelf, but we need actions to start taking place right now. That is our focus to move things forward on actions. We may not be focusing on research right now but it does not mean we will not be doing research. We just want to implement some of the results from previous research before moving on.

4 Structured Decision-Making is a process designed to standardize decision-making for managers, allowing them the ability to explore potential options and model potential outcomes.

In the years since, establishing more regulatory authority has become a top priority for SEFCRI, indicating a departure from data collection toward more active management. These priorities are detailed in the Florida Department of Environmental Protection's *Coral Reef Conservation Program 2020-2025 Strategic Plan.*

Despite SEFCRI's focus on data collection, one major challenge remains, the need for a no-take protected area, "We have been talking about all of these unknowns, but a very powerful management action already exists. We need areas offshore where we don't extract resources. We need areas where you can't remove fish or [substrate] for beach nourishment." In this case, while SEFCRI's Technical Advisory Committee acknowledges that no-take MPAs are practical and perhaps even necessary, the work being done has not yet incorporated them. The closest thing to a no-take MPA was the creation of the Kristen Jacobs Coral Reef Ecosystem Conservation Area, which was designated in 2018 (putting boundaries around the effective working area of SEFCRI). Plans are in the works for the eventual development of a new management plan with no-take zones (The Nature Conservancy, 2021). Nevertheless, SEFCRI does not operate with the same regulatory authority found within the FKNMS, and the timeline for creating these types of rules remains uncertain. It may come as a surprise to the reader that miles of sensitive reef habitat do not fall under a no-take MPA, but rather a patchwork of fisheries regulations.

We argue that many of these differences (e.g., active management versus preparatory management) can be explained because SEFCRI is newer. SEFCRI was established 13 years after the creation of the FKNMS and its actors first had to establish clear lines of communication and collect baseline data ("House Natural Resources," 2007). This means that the development and implementation of their Local Action Strategy is in different stages than the FKNMS Management Plan and its subsequent revisions. In addition, accelerated ecological degradation has caused many decision-makers to note the need for additional regulatory decision-making power, namely the creation of a no-take zone within the MPA.

Buffering: Are we really preparing for an uncertain future under climate change?

Given what we know now about SEFCRI's preparatory/research approach to Administrative Competence, it may not come as a surprise that SEFCRI placed more importance on Buffering (i.e., being prepared for the future) than the FKNMS. Research on novel ecosystems and coral reefs in the Anthropocene often highlights the need to be prepared for uncertainty while responding to environmental concerns (Good & Bahr, 2021; Hughes et al.,

2017a). Our research adds to this conversation, finding that SEFCRI does this through rich data collection and incorporating more novel considerations into recommendations via their Technical Advisory Committee. In this way, the concepts of Administrative Competence and Buffering were closely related for SEFCRI. For example, one interview respondent mentioned how data collected through SEFCRI was contributing to future coral reef restoration efforts, "We've learned a lot about restoration; we've harvested tissue to ensure genetic diversity and healthy populations. So, that's my hope and outlook on the future of the reef tract—getting it into a position where it can be restored." This quote highlights how preparing for the future is also an example of effective conservation.

SEFCRI decision-makers often appeared to be in the process of collecting novel types of data and implementing new projects. This novel data collection has been an essential task for SEFCRI because much of the northern reaches of the Florida Reef Tract had not been mapped until recently. For example, a press release from 2012 mentioned this type of novel research:

> *The Florida Department of Environmental Protection's Coastal and Aquatic Managed Areas' Coral Program completed its portion of data collection dives on more than 250 sites spread across 110 miles of the Florida Reef Tract that runs north from Biscayne National Park in Miami-Dade County to the St. Lucie Inlet in Martin County. The completion of this new survey on the northern portion of the reef marks the first time the entire Florida Reef Tract has been documented using the same method.* (Florida Department of Environmental Protection, 2012)

SEFCRI decision-makers also often mentioned Buffering in relation to public outreach and curbing the environmental impacts of dense population centers like Miami and Fort Lauderdale, which have more direct links to water quality issues to consider like surface runoff, drainage outflows and other anthropogenic point source pollutants (many of which are regionally distinct from the Florida Keys). For example, one of our interview respondents from SEFCRI's Technical Advisory Committee had this to say:

> *[You] have to accept that there are going to be different pressures [...] I don't know what the population is in the Florida Keys, but setting aside the tourists, it's tiny in comparison to the population [of Miami-Dade], so there are different pressures from that point of view [...] From population comes water quality, comes fishing pressure, comes diving pressure as well.*

In some cases, the concept of Buffering was defined similarly across both organizations. Concerning future threats, both organizations mentioned climate change, new species interactions, coral disease, maintaining sufficient genetic diversity and water quality. But, for the FKNMS, these concerns were often paired with suggestions for new zoning, new management proposals and new regulations—in other words new actions, which rely more heavily on their statutory power and completed management plans.

Some unique topics for the FKNMS included issues like emerging areas of concentrated visitor use in the Florida Keys, the increasing salinity of the Florida Bay, and fisheries management. Another difference was that the FKNMS could incorporate more past projects conducted within the Sanctuary to inform future management decisions. For example, take this quote from the director of the FKNMS, "We have some new science and condition reports to show what's working and where changes may need to be made."

Despite these differences, coral reefs and the threats they face are dynamic. They require anticipatory management, which carefully considers and incorporates contemporary research into an array of complementary management solutions, such as active coral restoration (a strength of the FKNMS) being coupled with robust data collection (a strength of SEFCRI) (Rogers et al., 2015).

Culture and norms: Do our policies reflect our cultural norms?

For the concept of Culture and Norms, each organization placed equal importance on this area, but there were distinct variations in their approaches. Public engagement and communication are critical to the success of projects that include ecological conservation, including the Florida Reef Tract (Druschke & Hychka, 2015; Sturmer, 2020). There is a difference between the value managers and the general public place on conservation. Through communication, transparency and awareness, the public can learn (via the help of managers) to promote conservation in their daily lives (Sturmer, 2020). Additionally, the inclusion of stakeholders within decision-making is critical for guiding management on the Reef Tract toward optimal outcomes inclusive of multiple values, interests and points of view (Rehr et al., 2012). Thus, public engagement and stakeholder involvement must always be a part of the adaptive management cycle. This cycle begins with a problem (e.g., coral reef degradation), prescribes management interventions (e.g., improving water quality) and communicates all successes and failures to the public (e.g., increasing transparency). In this way, stakeholders and the general public stay consistently involved (Druschke & Hychka, 2015).

However, when it comes to Culture and Norms, while there was crossover, the FKNMS focused more heavily on stakeholder *inclusion*, and SEFCRI focused more heavily on public *outreach*. In the FKNMS, stakeholder inclusion occurred whenever decision-makers were proposing to make changes to the management plan. They would solicit public comment through planned sessions and town-hall style meetings. Sanctuary Advisory Council meetings were also held to elicit the preferences of local stakeholders, such as fishers who were likely to be impacted by new rules and regulations. For example, take this representative quote from an economist with the NOAA Office of Marine Sanctuaries:

> *Most commercial fishermen cited the public process that resulted in the creation of the no-take Tortugas Ecological Reserve within the [FKNMS] as the reason for changing their attitudes. That process really allowed the local community to have a say in the Sanctuary boundaries and regulations.*

This demonstrates how public outreach is different from stakeholder inclusion. Public outreach is defined as public education, while stakeholder engagement is defined as stakeholders having their opinions considered in the decision-making process. It is also important to note a core difference between the Sanctuary Advisory Council (FKNMS) and the Technical Advisory Committee (SEFCRI)—the Sanctuary Advisory Council includes representatives from the fishing, diving and tourism industries (i.e., it is specifically geared toward stakeholder inclusion), while the Technical Advisory Committee is composed of research scientists from participating universities. A representative quote demonstrates the importance of stakeholder involvement for the FKNMS, where competing interests can find common ground through a participatory process and move management decisions toward shared interests:

> *Do I agree with him all the time? He knows I don't, and I know he doesn't [agree with me], but we're friends, and I can talk to him. So, those are things that I think you accomplish through finding [common ground], you find ways to work with [people], and it takes time, and that's part of the process.*

On the other hand, SEFCRI focused more heavily on public outreach. Decision-makers in SEFCRI focused more on public-service announcements, education campaigns, engagement with public figures (e.g., celebrities like Guy Harvey) and raising awareness about the Florida Reef Tract. For example, the success of the *Our Florida Reefs* initiative, a community-led campaign spearheaded by SEFCRI, was often mentioned. The following quote from a newspaper article represents an instance where public outreach was emphasized by leaders from the Florida Department of Environmental Protection's Coral Reef Conservation Program:

> *We are thrilled to have [conservation celebrity] Philippe Cousteau lend his voice to the Our Florida Reefs' campaign [...] It was also an honor for him to use [our] vessel as a platform to visit a thriving staghorn coral reef. [Cousteau's] passion [...] shines through in his [public service announcement] and segments of the coral reef education episode filmed that day.*

That is not to say that FKNMS ignored public outreach nor that SEFCRI ignored stakeholder inclusion. Both organizations emphasized stakeholder inclusion *and* public outreach, but to varying extents. For example, a NOAA representative at a SEFCRI Technical Advisory Committee meeting recommended, "Find common ground; projects or initiatives that stakeholders can work on together. Give them responsibility for conservation in [the] SEFCRI region. Focus on conservation initiatives that don't necessarily impact them [...]" In sum, while public outreach stood out for SEFCRI, stakeholder inclusion was certainly not ignored.

We suspect that the main reason for the differences in emphasis lies in the Administrative Competence context of SEFCRI and the FKNMS. While SEFCRI focused more on public engagement, the SEFCRI Region exists in an area where the population in 2020 was approximately 6.3 million, and its mandate was predominantly voluntary and research focused. In contrast, the population of Monroe county (i.e., the location of the FKNMS) was approximately 83,000 (U.S. Census Bureau, 2020). Therefore, engaging the public, who may not have equal access to coral reefs, might be more of a priority in urban centers. In contrast, engaging stakeholders might be a bigger priority in smaller populations where livelihoods are often inexorably tied to the tourism and fishing industries. Likewise, since decisions are part of the law in FKNMS, those impacted by laws must be consulted to ensure compliance and buy-in.

Issue framing: Have we considered the ecosystem before making decisions?

Issue Framing addresses whether an organization properly considers the ecological context before making management decisions. For example, are things like novel species compositions (e.g. the loss of stony coral) considered parts of the novel ecosystem context, or are previous baselines being used that may no longer be relevant? Our data generated information and quotes from both organizations, which indicate that adaptive governance is indeed occurring and acknowledging anthropogenic threats to the Florida Reef Tract. With the concept adequately considered by both organizations.

Topics related to Issue Framing were similar between the SEFCRI and the FKNMS data. Both institutions acknowledged significant stressors such as climate change, coral bleaching, ocean acidification, water quality, coral disease, new species interactions, coastal development and shifting baselines. All such stressors and issues were classified as requiring novel management solutions. Additionally, interview data from both organizations suggested there is a strong and growing sentiment that both MPAs are part of one novel ecosystem, and that the Florida Reef Tract has experienced irreversible changes. For example, take this quote from an interview with an FKNMS affiliated researcher:

> *Since I started my PhD and you can see the community changing on our reefs [...] after stony coral tissue loss disease went through, it's 100% a novel ecosystem in that it's relatively difficult to find living scleractinian corals [...] it's not only architecturally [different] in terms of coral cover, but also the [fish and invertebrate] community has changed dramatically [...] Hydrodynamically, biologically, ecologically; the ecosystem is absolutely one that is novel.*

A similar data point from an interview with a representative from SEFCRI's Technical Advisory Committee demonstrates the same perception:

> *I don't know that conditions will ever be ideal again. I don't think there's any going back to the way things were because the Keys, Miami Dade and Broward, they're not all the same as they were in the 1970s. The Keys may be a little better, because the majority of them are off septic now, they're all on a sewer system, but all the other areas there's not - there's no returning, [there] will never be a pristine ecosystem ever again. The Reefs are in their new normal now.*

Finally, consider the following quote from the director of the Office of National Marine Sanctuaries:

> *The world has changed and is changing in terms of what people are trying to do and what is happening to [ecosystems] everywhere. There is a lot of information on this topic. The [Florida Reef Tract] is a different place today. It is important to think about the long term picture. Thirty years ago it would have been hard to know that the ecosystem would deteriorate so much even when it was being managed. This is why it is so important to think differently in [the Florida Keys National Marine Sanctuary].*

Both institutions acknowledged the same threats and the need to manage these threats via conventional management strategies. But, there were differences between the organizations as well. SEFCRI consistently framed the novel ecology focusing on climate change, coral disease and new species interactions. In other words, SEFCRI looked at the broader ecological context to plan the best management practices. For instance, while acknowledging the effects of anthropogenic threats, this quote from a newspaper article demonstrates how SEFCRI's use of data collection helped to frame these changes:

> *The research team's project, funded by the Florida Department of Environmental Protection through the Southeast Florida Coral Reef Initiative, aims to identify the chain of causality land-based pollutants have rendered on the responses of reef-building coral and the health of coral reef communities in the South Florida Watershed.*

Alternatively, the FKNMS used existing regulations, management plans and results to iteratively propose changes to *existing* management. For example, the following quote shows an evolving process of management, where in response to anthropogenic threats, program priorities were being reevaluated to identify new opportunities for new management responses:

> *The Water Quality Protection Program recognizes the continuing water quality issues in [the] FKNMS and has committed to reevaluating the program's priorities and identifying new opportunities to more strategically pursue water quality improvements in the future [...] more detailed action plans will be developed and implemented to further advance the protection and restoration of [the Florida Keys'] water quality and associated marine resources.*

This quote demonstrates how FKNMS managers were implementing a new water quality program which is a 20-year collaborative effort to improve water quality for seagrass, coral reefs, fisheries and recreational access areas. This program was enacted to respond to water quality concerns which have been a threat to coral reefs in the Florida Keys for decades. Linked to the theme of Administrative Competence, FKNMS focus on adaptive governance in the face of a novel ecosystem is indicative of the regulatory, hierarchical way that MPA is set up, compared to the more voluntary structure of SEFCRI.

Power and authority: Does the public support decision-makers?

This concept is fundamental to adaptive environmental governance; power and authority means that managers have the ability to maintain public and political support while meeting objectives (Clement & Standish, 2018). Rules and regulations can negatively impact stakeholders, such as no-take zones, catch limits and boating restrictions. These limitations have the potential to become contentious among stakeholders whose livelihoods depend upon access to natural resources and may be affected by the inability to fish, dive, trap or otherwise engage with the ecosystem (Novoa et al., 2016). Regardless of how structured or fair a decision-making process may be, cooperation and compliance among individual stakeholders is key to the success of environmental governance. As a result, public compliance may falter or decline, reducing cooperation with law enforcement or respect for the institution (Maxwell & Maxwell, 2022; Wynveen et al., 2013).

Enforcement becomes problematic without citizens' cooperation, and conservation efforts may be undermined by harmful, negligent or criminal activity (Campbell et al., 2012; Maxwell & Maxwell, 2022). Research has shown that trust in formal government is not the only factor; social pressures and ethical values also influence individual cooperation, underpinning the success of environmental governance (Dalton et al., 2015; Maxwell & Maxwell, 2022). Given this information, it is easier to see how public perceptions of SEFCRI and the FKNMS play a critical role, enabling managers to do their jobs effectively (i.e., exercise Administrative Competence). Through being able to do their jobs well, managers are also able to incorporate additional novel ecosystem concepts (i.e., Buffering or Issue Framing) into their management strategies.

Given SEFCRI's largely informal nature, there was not as much regulatory implementation or enforcement. Instead, SEFCRI focused on the leadership roles of various parts of their administration and, once again, tended to focus more heavily on adequate preparation. For example, the potential for new marine zoning initiatives such as no-take zones, new forms of cooperative management and improvements via institutional changes were most important. For instance, consider this quote from a newspaper article:

> *The project [developed by the Southeast Florida Coral Reef Initiative (SEFCRI) Maritime Industry and Coastal Construction Impacts team] has identified [opportunities] for agencies to evaluate and encourage more effective coral reef permitting in Florida. 'We are developing scientific and legal support with our partners for this initiative,' said [a source].*

In other words, SEFCRI placed more importance on the ability to implement effective management when the resources become available in the future. This is not to say that they are not participating in active management, only that they emphasize having robust data, plans and mechanisms in place to act quickly. A representative data point that highlights SEFCRI's preparation for management actions, rules and regulations can be found in the following quote from SEFCRI's Technical Advisory Committee:

> *What happens to recommendations? [It's] not a management process, it is identifying recommendations that the committee will make to the SEFCRI chair and those will be forwarded to appropriate management agencies [...] Those will be brought forward and then have to go through their particular rule making processes. [The] Committee will provide recommendations to appropriate management agencies [who] will decide what to take forward into their work plans and rule-making process.*

For the FKNMS, Power and Authority was defined by measures to implement regulatory enforcement and procedural changes to existing management. In congruence with our other findings, these aspects focused on direct actions to enforce laws and regulations and ensure compliance within protected areas. For example, a representative from the Florida Fish and Wildlife Conservation Commission's Division of Law Enforcement had this to say:

> *FWC had seven offshore patrols [...] now five boats are going out to the Tortugas. Every time they go out there, they [enforce rules on somebody breaking them]. The goal is to make the stops and to find people in compliance, but they are not finding that to be true. This points to the notion that poaching may be taking place when the patrols are not there. They will continue patrolling the area and making cases.*

This quote, like many others, points to the increased enforcement capabilities of the FKNMS as a federal institution and highlights how the FKNMS can take direct action against any person(s) violating regulations within the sanctuary. However, it also highlights a concerning question: Is there enough law enforcement on the Florida Reef Tract? This question presents a segue into the criticisms found within our case studies. By examining criticisms in conjunction with instances of each concept, we understood more about what novel concepts were difficult to incorporate and why.

Criticisms of SEFCRI and FKNMS

A lack of enforcement

Enforcement was an issue of great concern for both organizations. For SEFCRI, this came down to regulatory enforcement not being a significant component of their management mandate. What enforcement does exist on the northern portions of the Florida Reef Tract seems to rely on statewide regulations rather than regulations designated explicitly for the SEFCRI Region. Therefore, criticism here mentioned the future need for regulated areas and enforcement capabilities. For instance, consider this quote from an interview respondent from the Florida Department of Environmental Protection:

> What's the point of having to manage a protected area if it in no way can be properly enforced? [...] I think that would be more of a possibility now, just in the way that funding has shifted, but [enforcement] is definitely a problem that would have to be addressed. That whatever additional regulation gets put [in the SEFCRI region], you have to have officers on the water to enforce it."

Alternatively, for the FKNMS, criticism of regulatory enforcement focused on public compliance and the ability of law-enforcement officers to do their jobs effectively within existing zones of the MPA. For example, one member of the Sanctuary Advisory Council mentioned the inability of law enforcement to adequately patrol the extent of the Reef Tract: "The law enforcement in the Sanctuary is [ineffective], and this is not about the individual law enforcement officers. I could go out there and break 90% of the rules, regulations, and laws and nobody would ever know it." This sort of criticism was supported by the Sanctuary Advisory Council, which at times noted that law enforcement was underfunded or lacked personnel:

> "The [Sanctuary Advisory Council] also voted to send a 'strong statement' to NOAA leadership, seeking an increase in sanctuary enforcement officers. The resolution notes that enforcement personnel for the Sanctuary's 2,900 square miles of water are 'at an all-time low, and constituents are demanding more enforcement resources.'"

However, despite these concerns, each organization actively works with the public to encourage voluntary environmental stewardship. For example, in the FKNMS, in addition to traditional law enforcement, the Sanctuary relies on "interpretive enforcement." Similar to SEFCRI's education and outreach, interpretive enforcement encourages voluntary compliance with regulations through education (Office of National Marine Sanctuaries, 2011). This strategy is also an example of how formal and informal institutions interact. A formal legal situation benefits via social norms, which help encourage compliance and accomplish the ultimate goal of conservation. An example of this occurs any time a dive charter full of tourists voluntarily reminds its guests that since they operate in a federal sanctuary, there is no taking, touching, kicking sand on the reef or harassing wildlife. This commonly occurs across many different dive charter operators in the FKNMS communities, it is voluntary, and it ensures guests know the rules of the sanctuary. Further institutional research has found that when differing institutions have similar conservation goals and are operational, these formal and informal institutions can complement each other (Osei-Tutu et al., 2015).

A lack of defined responsibilities

Another concern focused on SEFCRI's decentralized nature and a lack of clearly defined roles and responsibilities for management. The following quote references the substantial length of time required for some tasks to be completed by SEFCRI. Respondents suggested that the sheer number of stakeholders and managers involved in SEFCRI results in decisions taking longer to be made, including decisions as seemingly innocuous as the name of the MPA:

> *You know, SEFCRI was formed in 2004 and we finally got a name for that northernmost section of the reef tract in 2019. [...] Does it really take that long? They created 400 national parks in the less time that it took to create a name for 100 miles of reef. And without [agency] support, no [stakeholders] are going to buy into anything [...] especially because the [Kristin Jacobs Coral Reef Ecosystem Conservation Area] is in state waters, any sort of decision that is made that would affect fisheries must have [agency] support as well—It's so complex, talk about [too many] cooks in the kitchen.*

Research has shown that among larger groups of actors with greater diversity, like those involved with SEFCRI, policy and decision-making typically generate more conflict, which may require different considerations to counteract (Lubell et al., 2020). Research has also found that in actor networks like SEFCRI, multiple challenges arise from differing interests, goals, tensions and other external influences which can hinder management. For example, consider this quote from a manager in the SEFCRI Region, noting that it is difficult to coordinate between so many different organizations:

> *In a way [conservation is] getting harder and harder to accomplish [...] We're just learning about spawning, and what the wildlife is doing, and how the wildlife interacts with their own habitat. Having that science and knowledge should inform our policy and all the work that I think that these agencies and nonprofits are prioritizing. When you look at the different players for the Florida Reef Tract you're swimming in acronym soup, it's difficult.*

The lack of centralized authority (beyond the Florida Department of Environmental Protection) in SEFCRI has led to inaction due to the sheer number of managers involved and a lack of clear direction. In this example, the respondent discussed the use of Local Action Strategy Projects in SEFCRI to argue that their success depended on individual organizations' desire to enact particular projects. This dependency resulted in fragmentation where otherwise there could be an overarching comprehensive plan:

> *The Local Action strategy plans [...] are very much dependent on individuals being inspired to do something with them [...] The individuals concerned with SEFCRI [...] they work for Florida Fish and Wildlife Commission, or NOAA, or the Florida Department of Environmental Protection, or they work for local governments; there are representatives from all the counties [...] So, there's local government, and state government, and some from federal government agencies like the people from NOAA. And so, a lot of this stuff [is] highly dependent on somebody, or some member, being really inspired to pick up one [of the] proposed management activities and run with it.*

Distant decision-makers

The FKNMS received criticism for its top-down, distant, cumbersome bureaucratic hierarchy. Especially in the early years, stakeholders felt that the federal government was overreaching and limiting people's access to natural resources, impairing their abilities to generate livelihoods or recreate. For example, recreational anglers noted in a newspaper article, "The problem is that once we lose [a fishing area], we've lost it forever. We fight [the FKNMS] tooth and nail because when we know we're done, we're done."

However, much of this sentiment was dispelled via later periods of intense effort aimed at stakeholder involvement and public outreach, strategies which exist to build adaptive capacity (i.e., buffering) through communication, planning and shared values (Ayala-Orozco et al., 2018). For instance, during the initial phases of the 1997 FKNMS Management Plan, proposed zoning was reduced to a fraction of its original extent in an effort to compromise with various fishing stakeholders ("CMC Praises," 1996). A review conducted by the Sanctuary in 2009 found that stakeholder perceptions improved. Among the fishing community, opposition to the Sanctuary decreased dramatically from 78.4 percent in 1996 to 42 percent in 2009 ("User Group Support," 2009). Other interview respondents agreed that both stakeholder involvement and public outreach had effectively dispelled these kinds of fears in the 2020s. They mentioned that learning took place in two directions: from managers to the public and vice versa:

> *There's a lot of policy or community action initiatives that are brought up in those Sanctuary Advisory Council meetings. [I get to learn] why the community thinks the*

> *way that they do, and then, how we can be representatives of our constituents within the Council [to] move initiatives forward [...] There's certainly pushback from the Community about certain aspects of the plan [...], but as a whole, I think the community embraces the conservation of [this] critical resource [...], so I think there's a lot of understanding, respect, and appreciation for the Sanctuary.*

The difficulties facing SEFCRI and the FKNMS were often a combination of internal and external influences. This idea—that external influences exert control over internal functions—is supported by the literature, which finds that the political mood set by the state plays a vital role in effective resource management (Morrison et al., 2020). Political influence can be exerted by people in positions of authority, such as elected officials, to try and alter the outcome of a project based on personal motivations (Banfield & Wilson, 2017). Research on bureaucracy shows that concerns over job security, budgets and program financing can influence the attitudes and behaviors of managers who are responsible for implementing adaptive governance. This relationship is important in examining institutions' successes and failures and adaptive management that contend with regional ecosystems (Dunning, 2017).

For example, politics, namely the whims of elected leaders impacting management, came into play for SEFCRI. In 2015, former Florida governor Rick Scott's administration unofficially recommended that the Florida Department of Environmental Protection refrain from using the terms "climate change" or "global warming," a story that received considerable media attention (Korten, 2015). Some interviews referenced this political mood, stating that the policies of Governor Scott had direct impacts on the SEFCRI team, such as making future funding and job security uncertain for managers. These instances were often mentioned by interview respondents, who remarked that they suspected that some failures to implement recommendations, such as those generated during stakeholder meetings, resulted from an inhospitable political climate. One interview respondent noted that after generating plans for public outreach during the Our Florida Reefs initiative, those plans were archived.[5] Another respondent noted, "I knew plenty of people who were absolutely scared [that] they were going to lose their jobs if they were too vocal, and so I think some of the management of SEFCRI were affected by that."

5 As a reminder, the Our Florida Reefs initiative is a community planning process, spearheaded by SEFCRI, which aims to increase community awareness and involvement with the reef. For more information, visit: https://ourfloridareefs.org/

These criticisms are not meant to be exhaustive. Rather, they simply highlight some of the most interesting concepts and how SEFCRI and the FKNMS are responding to these issues. While criticisms make up a small portion of our data and are not wholly representative, they are informative and provide preliminary insight into areas that managers could focus on for future research or incorporate into decision-making.

Conclusion

In this chapter, we uncovered two distinct institutional arrangements which underpin the management strategies of SEFCRI and the FKNMS. Overall, we found that Administrative Competence was the most important part of managing a novel ecosystem for both the FKNMS and SEFCRI, indicating that both organizations value the preservation of the Florida Reef Tract. Administrative Competence reflects how important the ultimate goal of conservation is to each organization and to the managers working within.

There is also a relationship between all of the novel ecosystem concepts. For example, preparation for uncertain ecological conditions requires high levels of competence, and high levels of competence require the inclusion of novel criteria (there is a pattern of reciprocity). However, despite neither organization ignoring any of the novel ecosystem components, there were clear differences in the ways they approached management.

Overall, SEFCRI's coral reef management style was characterized by informal institutions, defined as management that included a large network of governmental and non-governmental actors (e.g., state agencies and non-profits, respectively). Ample opportunities for stakeholders to be involved in management were present in SEFCRI. This management style lacked rule-making authority, however. Instead, SEFCRI incorporated contemporary issues impacting the reef into their management considerations, focusing on future planning via research and generating new projects aimed at reducing ecological uncertainty. Specific activities to reduce uncertainty included generating knowledge about the northern reaches of the Florida Reef Tract, which until recently was largely unknown and informally managed, if at all. SEFCRI also focused heavily on broad public outreach initiatives to improve awareness of and appreciation for the northern reaches of the Florida Reef Tract. Finally, SEFCRI emphasized leadership roles among many actors, capitalizing on its decentralized, collaborative approach to management, deriving power from multiple actors within its network.

Alternatively, we found the FKNMS to be a formal management institution, defined by its top-down, authoritative approach to management, which relied heavily on governmental actors, laws and regulations. Given this formal structure, regulatory enforcement and procedural changes were common and important. The FKNMS used existing regulations, management plans and results to iteratively put the novel ecosystem concepts into practice. Past projects and tangible results were also used to inform future management decisions rather than collecting information for the first time. The FKNMS also strongly emphasized stakeholder engagement rather than public outreach, meaning that the FKNMS focused on a narrower group of actors whose livelihoods depend on the reefs of the Florida Keys. Ultimately, the FKNMS was characterized by regulatory actions, centralized leadership and active management concerns (i.e., concerns related to marine zoning and regulatory enforcement versus concerns associated with planning for future interventions).

This research is intended to be used as a stepping stone in novel ecosystem literature, generating new opportunities for advancement and research. Before our research, there was limited understanding of how these organizations function and adapt to the new environmental realities under climate change. Therefore, this research supplements the current literature on coral reef management in southeastern Florida and may help inform the management of similarly threatened ecosystems worldwide. As new management plans are developed, the capabilities and institutional arrangements of the FKNMS and SEFCRI will evolve. It remains to be seen what types of institutional arrangements will ultimately have the greatest impact on conserving the Florida Reef Tract, but significant progress is undoubtedly being made.

References

Allen, C. R., & Gunderson, L. H. (2011). Pathology and failure in the design and implementation of adaptive management. *Journal of Environmental Management, 92*(5), 1379–1384. https://doi.org/10.1016/j.jenvman.2010.10.063

Alvarez-Filip, L., Carricart-Ganivet, J. P., Horta-Puga, G., & Iglesias-Prieto, R. (2013). Shifts in coral-assemblage composition do not ensure persistence of reef functionality. *Scientific Reports, 3*(1), 1–5.

Anthony, K. R., Helmstedt, K. J., Bay, L. K., Fidelman, P., Hussey, K. E., Lundgren, P., Mead, D., McLeod, I. M., Mumby, P. J., Newlands, M., & Schaffelke, B. (2020). Interventions to help coral reefs under global change—A complex decision challenge. *PloS One, 15*(8), e0236399.

Ayala-Orozco, B., Rosell, J. A., Merçon, J., Bueno, I., Alatorre-Frenk, G., Langle-Flores, A., & Lobato, A. (2018). Challenges and strategies in place-based multi-stakeholder collaboration for sustainability: Learning from experiences in the Global South. *Sustainability, 10*(9), 3217.

Baker, B. (1999). First aid for an ailing reef: Research in the Florida keys national marine sanctuary. *BioScience, 49*(3), 173–178.

Banfield, E. C., & Wilson, J. Q. (2017). *Political influence* (pp. 1–10). Routledge.

Barrow, E., & Murphree, M. (2001). Community conservation: From concept to practice. In D. Hulme & M. Murphree, (Eds.), *African wildlife and livelihoods: The promise and performance of community conservation*, James Currey Ltd. (pp. 24–37).

Bodmer, M. D. V., Wheeler, P. M., Anand, P., Cameron, S. E., Hintikka, S., Cai, W., Borcsok, A. O., & Exton, D. A., (2021). The ecological importance of habitat complexity to the Caribbean coral reef herbivore Diadema antillarum: Three lines of evidence. *Scientific Reports*, *11*(1), 1–13.

Borrini-Feyerabend, G., & Hill, R. (2015). Governance for the conservation of nature. In G. L. Worboys, M. Lockwood, A. Kothari, S. Feary, & I. Pulsford (Eds.), *Protected area governance and management* (pp. 169–206). ANU Press.

Braverman, I. (2020). Shifting baselines in coral conservation. *Environment and Planning E: Nature and Space*, *3*(1), 20–39.

Burman, S., Aronson, R., & van Woesik, R. (2012). Biotic homogenization of coral assemblages along the Florida reef tract. *Marine Ecology Progress Series*, *467*, 89–96. https://doi.org/10.3354/meps09950

Campbell, S. J., Hoey, A. S., Maynard, J., Kartawijaya, T., Cinner, J., Graham, N. A., & Baird, A. H. (2012). Weak compliance undermines the success of no-take zones in a large government-controlled marine protected area. *PloS One*, *7*(11), e50074.

Carlisle, K., & Gruby, R. L. (2019). Polycentric systems of governance: A theoretical model for the commons. *Policy Studies Journal*, *47*(4), 927–952.

Chinn, H. (2020, November 19). "Dire outlook": Scientists say Florida reefs have lost nearly 98% of coral. *The Guardian*. https://www.theguardian.com/environment/2020/nov/18/coral-reefs-florida-dire-outlook

Cinner, J. E., McClanahan, T. R., Daw, T. M., Graham, N. A., Maina, J., Wilson, S. K., & Hughes, T. P. (2009). Linking social and ecological systems to sustain coral reef fisheries. *Current Biology*, *19*(3), 206–212.

Clement, S., Moore, S. A., & Lockwood, M. (2016). Letting the managers manage: Analyzing capacity to conserve biodiversity in a cross-border protected area network. *Ecology and Society*, *21*(3), 36–45.

Clement, S., & Standish, R. J. (2018). Novel ecosystems: Governance and conservation in the age of the Anthropocene. *Journal of Environmental Management*, *208*, 36–45. https://doi.org/10.1016/j.jenvman.2017.12.013

CMC Praises Florida's Sanctuary Management Plan. (1996, September 25). U.S. Newswire. https://advance.lexis.com/api/document?collection=news&id=urn:contentItem:3TWY-1SD0-0001-T0WH-00000-00&context=1516831

Cramer, K. L., Jackson, J. B., Donovan, M. K., Greenstein, B. J., Korpanty, C. A., Cook, G. M., & Pandolfi, J. M. (2020). Widespread loss of Caribbean acroporid corals was underway before coral bleaching and disease outbreaks. *Science Advances*, *6*(17), 1–10.

Dalton, T., Pollnac, R., & Forrester, G. (2015). Investigating causal pathways linking site-level characteristics, compliance, and ecological performance in Caribbean MPAs. *Coastal Management*, *43*(3), 329–341.

Douglas, A. (2020). Coral reef case study: Comparing current international and domestic coral reef laws and diving into the legal implications of countries failing to adequately protect and conserve these ecosystems as climate change worsens. *South Carolina Journal of International Law and Business*, *16*(3), 9.

Druschke, C. G., & Hychka, K. C. (2015). Manager perspectives on communication and public engagement in ecological restoration project success. *Ecology and Society*, *20*(1).

Dunning, K. H. (2017). Missing the trees for the forest? Bottom-up policy implementation and adaptive management in the US natural resource bureaucracy. *Journal of Environmental Planning and Management, 60*(6), 1036–1055.

Dunning, K. H. (2021). Adaptive Governance: The proposed port expansion in the Cayman Islands and its impacts to coral reefs. *Marine Policy, 124*, 104248.

Dustan, P. (2003). Ecological perspective: The decline of Carysfort Reef, Key Largo, Florida 1975–2000. *Ecological Forecasting: New Tools for Coastal and Ecosystem Management, NOAA Technical Memorandum NOS NCCOS*, 37–43.

Florida Department of Environmental Protection. (n.d.). Southeast Florida coral reef initiative technical advisory committee. Retrieved April 15, 2021, from https://floridadep.gov/rcp/coral/content/southeast-florida-coral-reef-initiative-technical-advisory-committee

Florida Department of Environmental Protection. (2004). Southeast Florida coral reef initiative: A local action strategy. https://ourfloridareefs.org/wp-content/uploads/2014/04/SEFCRI_LAS_FINAL_20May05.pdf

Florida Department of Environmental Protection. (2012, November 13). Coastal and aquatic managed areas' coral program completes reef fish survey [Press release]. https://content.govdelivery.com/accounts/FLDEP/bulletins/5be1fe

Florida Department of Environmental Protection. (2022). SEFCRI project reports and products. Retrieved February 25, 2022, from https://floridadep.gov/rcp/coral/content/sefcri-project-reports-and-products

Florida Keys National Marine Sanctuary. (n.d.b). Florida keys national marine sanctuary - Sanctuary advisory council. Retrieved September 23, 2021, from https://floridakeys.noaa.gov/sac/

Folke, C., Hahn, T., Olsson, P., & Norberg, J. (2005). Adaptive governance of social-ecological systems. *Annual Review of Environment and Resources, 30*, 441–473.

Frys, C., Saint-Amand, A., Le Hénaff, M., Figueiredo, J., Kuba, A., Walker, B., … Hanert, E. (2020). Fine-scale coral connectivity pathways in the Florida reef tract: Implications for conservation and restoration. *Frontiers in Marine Science, 7.* https://doi.org/10.3389/fmars.2020.00312

Gee, K., Kannen, A., Adlam, R., Brooks, C., Chapman, M., Cormier, R., Fischer, C., Fletcher, S., Gubbins, M., Shucksmith, R., & Shellock, R. (2017). Identifying culturally significant areas for marine spatial planning. *Ocean & Coastal Management, 136*, 139–147. https://doi.org/10.1016/j.ocecoaman.2016.11.026

Graham, N. A., Bellwood, D. R., Cinner, J. E., Hughes, T. P., Norström, A. V., & Nyström, M. (2013). Managing resilience to reverse phase shifts in coral reefs. *Frontiers in Ecology and the Environment, 11*(10), 541–548.

Graham, N. A., Cinner, J. E., Norström, A. V., & Nyström, M. (2014). Coral reefs as novel ecosystems: Embracing new futures. *Current Opinion in Environmental Sustainability, 7*, 9–14.

Graham, N. A., & Nash, K. L. (2013). The importance of structural complexity in coral reef ecosystems. *Coral Reefs, 32*(2), 315–326.

Gregg, K. (2013). Management considerations for the Southeast Florida coral reef ecosystem. Retrieved from https://floridadep.gov/sites/default/files/FDOU_Management_Considerations_0.pdf

Gershman, D., Elliott, C., Wondolleck, J., & Yaffee, S. (2012). *Florida Keys Natl. Marine Sanctuary.* Marine Ecosystem Based Management in Practice. Retrieved April 28, 2021, from http://webservices.itcs.umich.edu/drupal/mebm/?q=node/66

Good, A. M., & Bahr, K. D. (2021). The coral conservation crisis: Interacting local and global stressors reduce reef resiliency and create challenges for conservation solutions. *SN Applied Sciences, 3*(3), 1–14.

Head, B. W., & Alford, J. (2015). Wicked problems: Implications for public policy and management. *Administration & Society, 47*(6), 711–739.

Hobbs, R. J., Higgs, E. S., & Hall, C. (2013). *Novel ecosystems: Intervening in the new ecological world order.* Hoboken, NJ: John Wiley & Sons.

Hobbs, R. J., Higgs, E., & Harris, J. A. (2009). Novel ecosystems: Implications for conservation and restoration. *Trends in Ecology & Evolution, 24*(11), 599–605. https://doi.org/10.1016/j.tree.2009.05.012

House Natural Resources, Fisheries, Wildlife, and Oceans, H.R. 1205, Chris Bergh, Florida Keys Program Director. (2007, March 6). Coral reef conservation. CQ congressional testimony. https://advance.lexis.com/api/document?collection=news&id=urn:contentItem:4N6F-KX10-0003-11FB-00000-00&context=1516831

Hughes, T. P., Barnes, M. L., Bellwood, D. R., Cinner, J. E., Cumming, G. S., Jackson, J. B., … Scheffer, M. (2017a). Coral reefs in the anthropocene. *Nature, 546*(7656), 82–90.

Hughes, T. P., Kerry, J. T., Álvarez-Noriega, M., Álvarez-Romero, J. G., Anderson, K. D., Baird, A. H., … Wilson, S. K. (2017b). Global warming and recurrent mass bleaching of corals. *Nature, 543*(7645), 373–377.

Ives, C. D., & Kendal, D. (2014). The role of social values in the management of ecological systems. *Journal of Environmental Management, 144*, 67–72.

Jackson, J. B. C., Donovan, M. K., Cramer, K. L., & Lam, V. V. (2014). *Status and trends of Caribbean coral reefs* (pp. 1970–2012). Global Coral Reef Monitoring Network, IUCN.

Kerrigan, K. (2017). *SEFCRI local action strategy (2017).* Southeast Florida Coral Reef Initiative (SEFCRI). https://floridadep.gov/rcp/coral/documents/sefcri-local-action-strategy-2017

Korten, T. (2015). In Florida, officials ban term "climate change". WLRN Miami | South Florida. https://www.wlrn.org/environment/2015-03-09/in-florida-officials-ban-term-climate-change

Kuffner, I. B., & Toth, L. T. (2016). A geological perspective on the degradation and conservation of western Atlantic coral reefs. *Conservation Biology, 30*(4), 706–715.

Lessios, H. A. (2016). The great Diadema antillarum die-off: 30 years later. *Annual Review of Marine Science, 8*, 267–283. https://doi.org/10.1146/annurev-marine-122414-033857

Loh, T. L., McMurray, S. E., Henkel, T. P., Vicente, J., & Pawlik, J. R. (2015). Indirect effects of overfishing on Caribbean reefs: Sponges overgrow reef-building corals. *PeerJ, 3*, e901.

Lubell, M., Mewhirter, J., & Berardo, R. (2020). The origins of conflict in polycentric governance systems. *Public Administration Review, 80*(2), 222–233.

Manzello, D. P. (2015). Rapid recent warming of coral reefs in the Florida keys. *Scientific Reports, 5*(1), 1–10.

Maxwell, S. R., & Maxwell, C. D. (2022). Ecology and criminology? Applying the tenets of procedural justice on compliance to environmental regulations. *Criminology & Criminal Justice, 22*(2), 199–216.

Maynard, J., Byrne, J., Kerrigan, K., Tracey, D., Bohnsack, K., Pagan, F., … Waddell, J. (2017). *Coral reef resilience to climate change in the Florida Reef Tract.* Florida DEP Technical Report. https://repository.library.noaa.gov/view/noaa/16973

McClenachan, L. (2009). Documenting loss of large trophy fish from the Florida keys with historical photographs. *Conservation Biology, 23*(3), 636–643.

McClenachan, L. (2013). Recreation and the "right to fish" movement: Anglers and ecological degradation in the Florida keys. *Environmental History, 18*(1), 76–87.

McManus, J. W., & Polsenberg, J. F. (2004). Coral–algal phase shifts on coral reefs: Ecological and environmental aspects. *Progress in Oceanography, 60*(2–4), 263–279.

Morrison, T. H., Adger, N., Barnett, J., Brown, K., Possingham, H., & Hughes, T. (2020). Advancing coral reef governance into the Anthropocene. *One Earth, 2*(1), 64–74.

National Oceanic and Atmospheric Administration. (2015, October 30). *History of Florida keys national marine sanctuary.* Florida Keys National Marine Sanctuary. Retrieved February 22, 2022, from https://floridakeys.noaa.gov/history.html?s=about#:%7E:text=On%20November%2016%2C%201990%2C%20President,miles%20of%20Florida%20Keys%20waters

National Oceanic and Atmospheric Administration. (2022, February 9). *Restoring seven iconic reefs: A mission to recover the coral reefs of the Florida keys.* Habitat Conservation. Retrieved February 21, 2022, from https://www.fisheries.noaa.gov/southeast/habitat-conservation/restoring-seven-iconic-reefs-mission-recover-coral-reefs-florida-keys

Novoa, A., Kaplan, H., Wilson, J. R., & Richardson, D. M. (2016). Resolving a prickly situation: Involving stakeholders in invasive cactus management in South Africa. *Environmental Management, 57*(5), 998–1008.

Nuñez, M., & Dimarco, R. (2010). Keystone species. In *Berkshire encyclopedia of sustainability.* Berkshire Publishing Group. Retrieved April 13, 2021, from https://www.oxfordreference.com/view/10.1093/acref/9780190622664.001.0001/acref-9780190622664-e-516.

Office of National Marine Sanctuaries. (2011). *Florida keys national marine sanctuary condition report 2011* (p. 62). U.S. Department of Commerce, National Oceanic and Atmospheric Administration, Office of National Marine Sanctuaries.

Office of National Marine Sanctuaries. (2019). *Draft environmental impact statement for Florida keys national marine sanctuary: A restoration blueprint.* U.S. Department of Commerce, National Oceanic and Atmospheric Administration, Office of National Marine Sanctuaries.

Osei-Tutu, P., Pregernig, M., & Pokorny, B. (2015). Interactions between formal and informal institutions in community, private and state forest contexts in Ghana. *Forest Policy and Economics, 54*, 26–35.

Pahl-Wostl, C. (2009). A conceptual framework for analysing adaptive capacity and multi-level learning processes in resource governance regimes. *Global Environmental Change, 19*(3), 354–365. https://doi.org/10.1016/j.gloenvcha.2009.06.001

Pahl-Wostl, C. (2015). *Water governance in the face of global change.* Springer International Publishing. https://doi.org/10.1007/978-3-319-21855-7

Pandolfi, J. M., Bradbury, R. H., Sala, E., Hughes, T. P., Bjorndal, K. A., Cooke, R. G., McArdle, D., McClenachan, L., Newman, M. J., Paredes, G. & Warner, R. R. (2003). Global trajectories of the long-term decline of coral reef ecosystems. *Science, 301*(5635), 955–958.

Perry, C. T., & Alvarez-Filip, L. (2019). Changing geo-ecological functions of coral reefs in the anthropocene. *Functional Ecology, 33*(6), 976–988.

Perry, C. T., Murphy, G. N., Graham, N. A., Wilson, S. K., Januchowski-Hartley, F. A., & East, H. K. (2015). Remote coral reefs can sustain high growth potential and may match future sea-level trends. *Scientific Reports, 5*(1), 1–8.

Pilnick, A. R., O'Neil, K. L., Moe, M., & Patterson, J. T. (2021). A novel system for intensive Diadema antillarum propagation as a step towards population enhancement. *Scientific Reports, 11*(1), 1–13.

Precht, W. F., & Miller, S. L. (2007). Ecological shifts along the Florida reef tract: The past as a key to the future. In R. B. Aronson (Ed.), *Geological approaches to coral reef ecology* (pp. 237–312). Springer. https://doi.org/10.1007/978-0-387-33537-7_9

Rehr, A. P., Small, M. J., Bradley, P., Fisher, W. S., Vega, A., Black, K., & Stockton, T. (2012). A decision support framework for science-based, multi-stakeholder deliberation: A coral reef example. *Environmental Management*, *50*(6), 1204–1218.

Reisewitz, A., & Harper, W. (2013). *Our Florida reefs strategic communications plan, 2013–2016.* Retrieved April 11, 2021 from https://floridadep.gov/rcp/coral/documents/our-florida-reefs-strategic-communications-plan-2013-2016

Rijke, J., Brown, R., Zevenbergen, C., Ashley, R., Farrelly, M., Morison, P., & van Herk, S. (2012). Fit-for-purpose governance: A framework to make adaptive governance operational. *Environmental Science & Policy*, *22*, 73–84.

Rogers, A., Harborne, A. R., Brown, C. J., Bozec, Y. M., Castro, C., Chollett, I., Hock, K., Knowland, C. A., Marshell, A., Ortiz, J. C. & Razak, T. (2015). Anticipative management for coral reef ecosystem services in the 21st century. *Global Change Biology*, *21*(2), 504–514.

Saint Petersburg Coastal and Marine Science Center. (2017). *Time series of Grecian rocks coral reef.* U.S. Geological Survey. Retrieved February 21, 2022, from https://www.usgs.gov/media/images/time-series-grecian-rocks-coral-reef

Sleasman, K. (2009). Coordination between Monroe County and the Florida keys national marine sanctuary (FKNMS). *Ocean & Coastal Management*, *52*(1), 69–75. https://doi.org/10.1016/j.ocecoaman.2008.09.002

Southeast Florida Coral Reef Initiative. (2019). *Who we are.* Retrieved November 2, 2020, from https://southeastfloridareefs.net/who-we-are/

Storlazzi, C. D., Reguero, B. G., Cole, A. D., Lowe, E., Shope, J. B., Gibbs, A. E., … Beck, M. W. (2019). *Rigorously valuing the role of US coral reefs in coastal hazard risk reduction (No. 2019–1027).* US Geological Survey. https://doi.org/10.3133/ofr20191027.

Sturmer, K. (2020). *Connecting the dots: Strengthening the communication between researchers, community members, and policy makers in the Florida reef tract.* Royal Road.

Thaker, J., Howe, P., Leiserowitz, A., & Maibach, E. (2019). Perceived collective efficacy and trust in government influence public engagement with climate change-related water conservation policies. *Environmental Communication*, *13*(5), 681–699.

The Nature Conservancy. (2021). *Resilience action plan for Florida's coral reef 2021–2026.* https://www.nature.org/content/dam/tnc/nature/en/documents/Resilience-Action-Plan-for-Floridas-Coral-Reef-2021-2026.pdf

Toth, L. T., Stathakopoulos, A., Kuffner, I. B., Ruzicka, R. R., Colella, M. A., & Shinn, E. A. (2019). The unprecedented loss of Florida's reef-building corals and the emergence of a novel coral-reef assemblage. *Ecology*, *100*(9). https://doi.org/10.1002/ecy.2781

U.S. Census Bureau. (2020). *U.S. Census Bureau QuickFacts: Monroe County, Florida; Martin County, Florida; Palm Beach County, Florida; Miami-Dade County, Florida; Broward County, Florida.* Census Bureau Quick Facts. Retrieved December 30, 2021, from https://www.census.gov/quickfacts/fact/table/monroecountyflorida,martincountyflorida,palmbeachcountyflorida,miamidadecountyflorida,browardcountyflorida/PST045221

U.S. Department of Commerce. (2007). *Florida keys national marine sanctuary revised management plan.* https://floridakeys.noaa.gov/mgmtplans/2007.html

User Group Support for Florida Keys National Marine Sanctuary Continues to Increase. (2009, June 4). Targeted News Service. https://advance.lexis.com/api/document

?collection=news&id=urn:contentItem:7VVR-GH51-2PW7-T0C6-00000-00&context=1516831

Wagner, D. E., Kramer, P., & Van Woesik, R. (2010). Species composition, habitat, and water quality influence coral bleaching in southern Florida. *Marine Ecology Progress Series, 408*, 65–78.

Walton, C. J., Hayes, N. K., & Gilliam, D. S. (2018). Impacts of a regional, multi-year, multi-species coral disease outbreak in Southeast Florida. *Frontiers in Marine Science, 5*, 323.

Weil, E., Smith, G., & Gil-Agudelo, D. L. (2006). Status and progress in coral reef disease research. *Diseases of Aquatic Organisms, 69*(1), 1–7.

Wynveen, C. J., Connally, W. D., & Kyle, G. T. (2013). Pro-environmental Behavior in Marine protected areas: The cases of the great barrier reef marine park and the Florida keys national marine sanctuary. *Journal of Park and Recreation Administration, 31*(2). https://js.sagamorepub.com/jpra/article/view/3884

Yates, K. K., Zawada, D. G., Arsenault, S. R., & Fehr, Z. W. (2019). *Seafloor elevation change from 2016 to 2017 at Crocker reef, Florida keys—Impacts from Hurricane Irma.* US Geological Survey Data Release.

Young, O. R. (2016). *On environmental governance: Sustainability, efficiency, and equity.* Routledge.

Chapter 6

HOW COMMUNITIES ARE ORGANIZING TO CONTEST MAJOR INFRASTRUCTURE PROJECTS THAT MAY DAMAGE CORAL REEFS: THE PORT OF MIAMI CASE

By Sabine Bailey, Daniel Morris, and Kelly Dunning

Imagine the hustle and bustle of downtown Miami, teeming with live music, filled with flavors of Latin American cuisine, aromas of Cuban coffee and tourists and locals alike enjoying miles of pristine white-sand beaches. A short distance to the south, lies the Biscayne Bay, which is home to several listed species under the most important wildlife conservation policy on the books, the Endangered Species Act, including manatees, crocodiles, several sea turtle species, small-tooth sawfish and staghorn coral (*Acropora cervicornis*). Just a couple miles down the road on U.S. Route 1, the heart of the coral reefs of South Florida, or the Florida Reef Tract can be found where it stretches 300 miles south to the Dry Tortugas National Park, making it the third largest barrier in the world. To the west of Miami, we find Everglades National Park, which is a designated World Heritage site with one of the most unique wetland ecosystems in the world. Everglades National Park is also one of the wettest regions in the United States, with many of its waterways finding their final destination in the Biscayne Bay, and ultimately into the Florida Reef Tract ecosystem, showing the interconnectedness of all of these South Florida ecosystems. These tropical coastal seascape ecosystems are characterized by seven connected ecosystems that include coastal strand, mangroves, seagrass beds, coral reefs, hard-bottom habitat, silt-bottom habitat and rocky-bottom habitat. Right in the middle of this mosaic of ecosystems lies the Port of Miami, nicknamed the "Gateway to the Americas" and "Cruise Capital of the World." In this chapter we explore how major infrastructure projects may degrade delicate and imperiled coastal ecosystems despite strict federal, state and local laws, and how people and interest groups are working to protect these ecosystems.

Figure 6.1 A cruise ship entering the Port of Miami. Courtesy of Felix Mizioznikov/ Shutterstock.

The Dredging History of the Port of Miami

To meet the demands of the burgeoning market, the Port of Miami has undergone many dredging projects to accommodate larger shipping vessels and cruise ships throughout its history.[1] Beginning in 1990, Congress authorized the deepening of the Port to 42 feet, which was completed in 1993. Since then, every 10 to 15 years, another port expansion dredging project has been implemented.

What makes this situation concerning is that threatened coral species, such as staghorn coral, exist in the shipping channel by the Port of Miami. Corals in this habitat live in large individual coral colonies and clusters and may not serve the same ecosystem services as barrier reefs, such as coastal protection and tourism. However, the corals have existence value as well as option value,[2] as they may be more resistant to coral disease or coral bleaching caused by warming waters (Udel, 2014). Every time ports are dredged, the surrounding environment is impacted, at the very least, temporarily, by plumes of sediment with potential long-term effects such as widespread coral mortality (Bak, 1978; Costa et al., 2014; Cunning et al., 2019; Valadez-Rocha & Ortiz-Lozano, 2013). As the Port of Miami is surrounded by vulnerable ecosystems, such as seagrass and coral habitat, both of which have strong environmental protection policies (outlined in Appendix 6), these major port infrastructure upgrades

1 Dredging is the removal of unwanted sediment and debris from the ocean floor. As ships are getting larger, they need deeper navigation channels to ensure safe passage to ports, which is accomplished through dredging (US Department of Commerce, n.d.).

2 Option value considers the value or benefit a potential resource may have in the future.

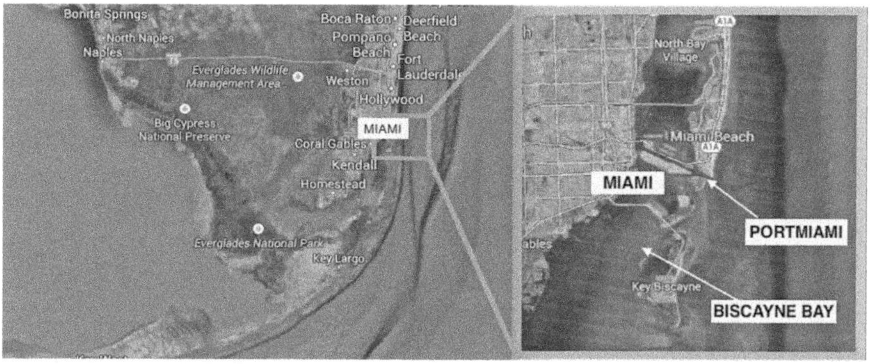

Figure 6.2 The Port of Miami known as "PortMiami" near Biscayne Bay, Florida. Adapted from Google Maps.

are highly controversial and worth investigation. The most recent dredging project for the Port of Miami, which began in 2012 and was completed in 2015, culminated in an environmental dispute which is still ongoing as of 2022.

This chapter explores how this infrastructure project unfolded and how communities organized to contest the project. Specifically, we look at the narratives surrounding the environmental dispute used by two distinct groups: those supportive of the project and those opposed to it. We argue that narratives played a major role in influencing policy actions during the Miami port project and dredging projects planned in Florida's future (Frazer et al., 2021). We examine how stakeholders from both sides of the dispute cast other stakeholders or entities as victims, villains and heroes and how highly or villainously they speak about each other. We also take an in-depth look at the use of science in narratives across both stakeholders groups because discovering the truth behind the science that was conducted and reported during the Port of Miami dredging project became the pinnacle of this dispute.

The following sections present our case study data, with methodology and data collection information contained in the Appendix for Chapter 5.

The Dispute of the 2015 Port of Miami Dredging Project

As part of the *We Can't Wait* initiative under the Obama administration, the 2015 Port of Miami dredging project was approved and fast-tracked in 2012.[3] The deep dredge project was in response to the widening of the Panama

3 The We Can't Wait Initiative was a campaign launched by the Obama Administration in 2011 to help revive the economy via executive orders and other administrative rule-making bypassing a gridlocked Congress.

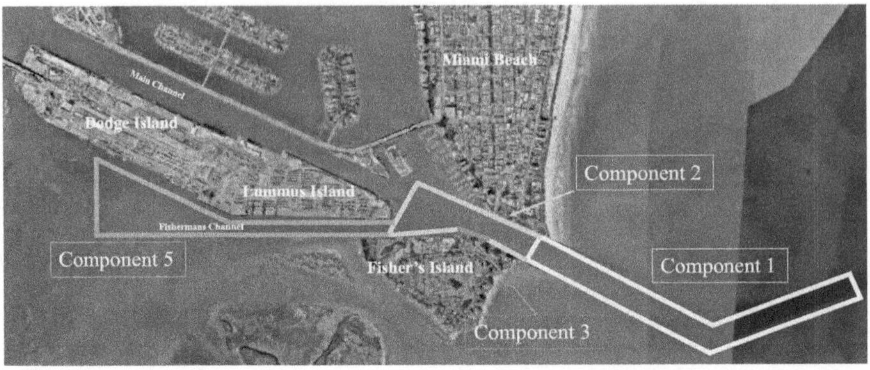

Figure 6.3 The Port of Miami phase III deepening project plan. Adapted from Dial Cordy and Associates Inc, 2016 using Google Earth.

Canal, which was completed in 2016, with similar port improvements occurring all over the world. The Port of Miami needed to be ready to accommodate and dock Post-Panamax size vessels,[4] which stakeholders who supported the project believed would become the industry standard. There were five dredging components to the project, which included deepening the main channel to 52 feet, deepening and widening a nearby channel, Fisherman's Channel, to 50 feet, widening the channel entrance by 300 feet, and expanding the turning basin for large ships near Lummus Island (see Figure 6.3).[5]

The U.S. Army Corp of Engineers (the USACE or "the Corps"), a federal agency, supervised the project. They contracted Great Lakes Dredge and Dock, a private company, for dredging and mitigation services. Great Lakes Dredge and Dock also contract multiple private companies to assist them in their tasks. The project was funded by federal, state and county dollars and cost $205.6 million. This was the first time a USACE project was not solely funded by federal dollars. The director of PortMiami at the time praised the former governor of Florida, Rick Scott, and former mayor of Miami-Dade County, Carlos Gimenez, for spearheading the project and contributing over

4 Post-panamax II vessels (2000-onward) can transport 6,000–8,500 TEUs, while neo-Panamax size vessels (2014-onward) can transport 12,500 TEUs (Rodrigue, 2020).

5 The following describes the dredging work associated with each component of the dredging project. Component 1: Widen seaward portion from 500 to 800 feet and deepen channel from 44 to 52 feet. Component 2: Increase Fisher Island turn basin from 1200 to 1500 feet and deepen from 42 feet to 50 feet. Component 3: Widen turn at Fisherman's channel and deepen to 50 feet. Component 5: Expand berthing area by 60 feet and deepen channel from 42 to 50 feet.

a million dollars to state and county economies. The dredging itself took two years to complete, beginning in November 2013 to September 2015.

During that time, over five million cubic yards of ocean bottom was removed (USACE, n.d.). Despite mitigation efforts, some people believed that extensive coral mortality occurred during the project. This is significant because Florida's reefs are estimated to have a value of $8.5 billion dollars annually from tourism, fisheries, local sales and jobs (Towle et al., 2020).

In 2008, the USACE and Environmental Protection Agency issued an ocean dumping permit to the USACE Jacksonville District that authorized the disposal of dredged materials to specific dumping areas. Due to the acknowledged environmental impact of the dredging project, the private firm Great Lakes Dredge and Dock would be responsible for both dredging and implementing mitigation efforts. Mitigation efforts included the restoration of 16.6 acres of seagrass beds, creation of 9.8 acres of artificial reefs and relocation of all hard corals greater than 25 cm and 1,300 hard corals between 10 cm and 25 cm onto artificial or existing reefs outside the designated impact area (USACE, 2014). In addition, they subcontracted another private firm, Dial Cordy and Associates, to conduct a baseline environmental assessment and monitor present and post-construction conditions of reef, seagrass and hard-bottom habitat.

In May 2010, subcontracted firm, Dial Cordy and Associates, identified 31 staghorn coral (*Acropora cervicornis*) colonies, within the estimated impact zone of 150 meters from the dredging site. There was a special focus on staghorn coral because, under the Endangered Species Act, consultation is required by the National Marine Fisheries Service to authorize any take of a listed species. Upon review of the initial survey in 2010, the National Marine Fisheries Service provided a Biological Opinion Statement and approved an Incidental Take Statement in 2011,[6] authorizing the USACE to take the 31 identified staghorn colonies and transplant them outside the impact zone.

---

6 A Biological Opinion is a document that states the National Marine Fisheries Service's opinion on whether the actions of Federal agencies are likely to destroy critical habitat or threaten the continued existence of listed species. If appropriate, this can be accompanied by an Incidental Take Statement that provides an exemption to "take," which can include injuring, wounding, hunting, killing or collecting a specific amount of the listed species to minimize impacts from Federal actions. A reinitiation of consultation by either party must be requested if the amount of take exceeds what is stated in the Incidental Take Statement, or if new information and conditions have changed that impact the listed species to an extent not previously considered (FWS, n.d.; NOAA Fisheries, n.d.).

It is only by following this specific process that private firms are allowed to damage coral species in the United States.

The Florida Department of Environmental Protection was also ready to issue its required permit to the USACE to proceed with the 2015 Port of Miami dredging project in August 2011.[7] Upon reviewing the proposed permit, a lawsuit was filed against the USACE and the Florida Department of Environmental Protection by non-governmental organizations and concerned citizens, including the Tropical Audubon Society, and Biscayne Bay Waterkeeper (a group known as the "Reef Guard").[8] The Reef Guard wanted to strengthen the mitigation efforts proposed in the permit, and they also scrutinized the impacts of confined blasting to deepen the channel.[9] The lawsuit was settled when the USACE and the Florida Department of Environmental Protection agreed to amend the permit to limit the daily blasting time frame and strengthen the monitoring parameters. Miami-Dade County also agreed to establish $1.3 million dollars in a trust fund,[10] which would go toward monitoring of relocated corals and seagrass restoration, as well as enhance other environmental projects in South Florida (Miller, 2012). In 2012, the permit was issued to the USACE to proceed with the project.

One month prior to dredging, in late September 2013, the subcontracting firm Dial Cordy and Associates went to identify the previously surveyed 31 staghorn coral colonies. However, due to miscommunication among organizations, it was found that there were actually over 243 staghorn coral colonies within the dredging impact zone.[11] The USACE attempted to consult with the National Marine Fisheries Service on how to proceed, given this error. However, a government shutdown from October 1–17, 2013, meant that all agencies were on furlough, and thus there was no formal procedure.

7 Permit No. 0305721-001-BI.

8 The Biscayne Bay Waterkeeper is now known as the Miami Waterkeeper.

9 Confined blasting is when a hole is drilled in rock or limestone, filled with explosives and capped with materials such as crushed rock to confine the explosion in the rock.

10 A trust fund is a collection of assets managed on the behalf of a third party. In this case, monies received to a county government trust fund are supervised by a county department to charge and collect feets to disburse money toward specific duties. For example, monies from the Biscayne Bay Environmental Enhancement Trust Fund can only be disbursed for environmental enhancement activities, such as the installation and management of artificial reefs and mooring buoys in the Biscayne Bay under the Board of County Commissioners (*Biscayne Bay Environmental Enhancement Trust Fund*, 2022).

11 According to Interview #4, there was an extrapolation error by another organization and 31 coral colonies should have been multiplied by nine to get the full number of coral colonies within the estimated dredging impact zone (~279 coral colonies).

Instead, 38 staghorn colonies were relocated and 26 permanent monitoring sites were set up by the subcontractor Dial Cordy and Associates and dredging started in November 2013. The difference in corals resulted in the loss of those not part of the original mitigation work.

In 2014, the public reported high turbidity and sedimentation in the water column, which became a major controversy. Concerning dredging, sedimentation is when ocean bottom materials from dredging, such as sand, are suspended in the water column appearing chalk white and milky in typically clear water. In response, the Miami-Dade County Division of Environmental Resources Management and Florida Department of Environmental Protection each sent teams of scientists to inspect the dredging site in July 2014. They both discovered that widespread disturbances to corals due to dredging-related activities were occurring well beyond what was allowed in the permit. The next month, the Florida Department of Environmental Protection sent the USACE a Warning Letter that requested the USACE to undertake an assessment of the sedimentation impact from dredging operations. From these inspections, the National Marine Fisheries Service requested an emergency relocation of all staghorn colonies in September 2014. The Reef Guard filed their second lawsuit in October 2014, because they believed the USACE was in violation of the Endangered Species Act.[12] They claimed that (1) the USACE had exceeded the amount of allowable take of staghorn corals as per the 2011 Incidental Take Statement, (2) continued dredging activities during consultation periods when it should have paused, and (3) that sedimentation from dredging was adversely impacting critical reef habitat. In response, the USACE agreed to pay over $400,000 to relocate staghorn coral colonies to a coral nursery supervised by the University of Miami.

The USACE was also reported by the EPA, which raised concerns that violations of its 2008 permit were taking place. Specifically, the EPA was alarmed that the scows, or vessels which transport the dredging material to designated dumpsites, were overflowing to save companies money and minimize the number of trips and cost. The scows passed over critical coral habitat, potentially risking their health, which was another violation, this time against the Marine Protection, Research and Sanctuaries Act. The EPA outlined an additional 125 events of noncompliance. The USACE denied these violations stating that they had monitoring devices that notified them whether scows were overflowing. Despite growing opposition from the public,

12 Plaintiffs include the Tropical Audubon Society, Biscayne Bay Waterkeeper, and Captain Dan Kipnis, from the first litigation, in addition to the Miami-Dade Reef Guard Association.

official warnings, violations of their permit and environmental policies, the USACE completed the dredging project in September 2015.

The Reef Guard reopened the 2014 lawsuit against the USACE for violating the Endangered Species Act in 2016. It was settled in 2018 which secured the restoration of an additional 10,000 corals in Miami-Dade county (*Miami Waterkeeper*, 2018). As per the 2012 deepening permit issued by the Florida Department of Environmental Protection, monitoring and impact assessments led by the subcontracting firm Dial Cordy and Associates were required for three years post dredging operations. They reported that 219 out of 612 coral colonies saved as part of the mitigation work died during the project, but they claimed that only seven of those colonies' deaths were associated with dredging. The report stated that the majority of coral death was associated with an outbreak of coral disease. Dial Cordy and Associates also noted that a coral bleaching event and novel disease outbreak made it more difficult to identify the source of coral mortality.[13]

The impacts of dredging on adjacent coral colonies are at the center of this environmental dispute. All stakeholders involved accepted that changes and improvements need to be implemented for the future. Incorporating the lessons learned from this project is especially important because multiple dredging projects are already planned for the future. For instance, further deepening of the Port of Miami was once again approved in September 2020. Additionally, 30 miles north of the Port of Miami, Port Everglades is in the preconstruction phase for a similar dredging project. However, the Reef Guard has already pressed legal actions to ensure mitigation efforts are sufficient, and along with the ongoing COVID-19 pandemic, the project continues to be delayed.

Narrative Strategies Used by Stakeholders

The Miami case presents a classic environmental dispute: Do we prioritize potential economic growth of the freight and tourism sectors over ecosystems?

13 When corals are exposed to adverse environmental conditions for prolonged periods of time, especially with rising sea surface temperatures, corals will respond by expelling the symbiotic zooxanthellae, revealing their underlying white calcium carbonate skeletons. This response is called "bleaching" (Wagner & Ylä-Anttila, 2018). It was identified that the first sightings of stony coral tissue loss disease were recorded during the Port of Miami dredging project (Precht et al., 2019). Presently, stony coral tissue loss disease is one of the biggest and widespread threats to coral reefs in the Caribbean and has resulted in devastating coral reef mortality (Estrada-Saldıvar et al., 2021; Muller et al., 2020).

Due to the values that reefs hold, the stringent environmental regulations, and the impacts on corals, the Miami dredging project garnered significant media attention where different interests used different narratives to push for policy changes. The project became the subject of an environmental dispute among stakeholders, namely those for the deepening project, claiming it must be done for economic growth and nationwide prestige, and those against it, claiming the health of the corals and environment faced significant risks.

We've unpacked the events that led to this environmental dispute. However, to further understand and analyze it, we look to the strategic narratives used by local stakeholders. We argue that these narratives are important because they lead to greater coral monitoring and relocation efforts as well as the delay of the Port Everglades dredging project, another dredging project 30 miles north of the Port of Miami, by at least five years.

Drawing from public policy literature, we use the Narrative Policy Framework (Shanahan et al., 2018) to examine and critically study how stakeholders employ three types of strategic narratives to push their policy preferences (either supportive of or opposed to the port project). We assess how stakeholders cast entities as victims, villains, heroes (*characters*), how they speak exceedingly highly or villainously of them (*devil-angel shift*) and the role of science and how they use science to justify or rationalize their narratives (*science*). Table A.6.2, A.6.3, and A.6.4 summarize data sources and qualitative coding findings in Appendix 6.

Victims, Villains and Heroes in Our Case Study

We examine how opponents and supporters of the 2015 Port of Miami dredging project cast others as three types of characters: victims, villains and heroes. Victims are perceived as people or entities that are harmed or suffer losses. Villains are perceived as policy actors that inflict harm and suffering. Heroes are perceived as policy actors that provide solutions to relieve the harm being done.

Opponents of the project cast the Biscayne Bay environment, coral reefs and citizens of Miami-Dade county as the victims. They believed that the dredging project would bring irreparable harm to the surrounding environment, especially impacting coral habitat in and beyond the Biscayne Bay. They also perceived Miami-Dade citizens as victims who would have to bear the economic burden (through taxes) to pay for any long-term consequences that may come from the dredging project. Below is a quote from one of the leaders of the opposing coalition, the Reef Guard:

> *We've seen profound and severe impacts to our reef just off of Miami…it looks like a moonscape…This damage stems from the fact that the Corps and the contractors simply weren't following the rules that were laid out for them when they started this project…I'm not quite sure that county taxpayers fully understand that they are on the hook for paying for this.* (Alvarez, 2015)

The quote above also shows that opponents cast the USACE and its contractors as the prominent villains for bringing the damage on the environment through violating permits and rules. Opponents cast themselves as heroes and underdogs for defending the environment, garnering public attention on the dispute and bringing legal actions against the USACE to enhance mitigation efforts. The quote below shows how the Reef Guard believed they were acting as the heroes:

> *We're doing our best to educate Congress on the risks of this project […] A lot of these things feel like they're too big to be dealt with on a local level. But avoiding impacts from dredging is something we can control on a local level.* (Upton, 2016)

Surprisingly, supporters of the project also cast the environment, and citizens as victims, though for different reasons. They acknowledged that there would inevitably be minimal impacts on the environment but were implementing mitigation efforts, including seagrass restoration and coral relocation to offset them. Supporters also believed that delaying the project would result in a cost to Miami-Dade taxpayers, stating that:

> *Any adjustment to the placed pilings would [delay the project and] […] cost local taxpayers thousands of dollars, to remobilize the contractor, purchase new piling, re-drive the new pilings, and make necessary engineering adjustments to the size of the site.* (USACE Jacksonville District Website, 2014)

For supporters, canceling the dredging project was not an option and would have resulted in a nationwide loss of economic opportunities, as is expressed in the quote below focusing on how the city of Miami would lose a competitive edge in global trade relations:

> *With the opening of an expanded Panama Canal in 2014, the Port of Miami, as the closest U.S. port to the Canal, is well positioned to drastically increase trade, but only if the port's channel is deepened to 50 feet to accommodate the new generation of larger cargo vessels. Failure to do so will send this business, and thousands of jobs, elsewhere – namely to foreign transshipment hubs. It is time for our elected leaders to put aside political differences and fund the Deep Dredge Project ensuring that the*

> *Port of Miami remains a competitive player in the global marketplace. Failing to do so would be a loss not only to our local and state economies, but to the entire nation.* (Gloor, 2010)

Supporters perceived villains as any entity delaying the prosperous dredging project and causing Miami to lose a competitive edge in global trade networks. Villains changed over the course of the project for supporters. When federal and state governments were negotiating how to dole out funds for the project in 2010 and 2011, the Floridian government (who supported the project) believed that the federal government was not financially supporting the project enough and portrayed them as villains. These disagreements likely led to the project being funded by not solely federal dollars but also state and county public dollars, a first for any USACE project. Once the project was approved and funded and when dredging began, supporters shifted the narrative, presenting opponents of the project as villains instead, as highlighted in the quote below:

> *Opponents of the dredge had employed "delay tactics" [...] [and] were afforded ample opportunity to have their voices heard. The economic impact to our community [was] too important to let a small group of obstructionists delay it any longer than necessary.* (McGrory and Curtis Morgan, 2012)

Supporters of the dredging project portrayed the dredging project itself, Floridian government officials and the USACE as heroes. They believed that the project allowed the Port of Miami to stay competitive in world trade and would bring economic prosperity not just to Florida but nationwide. Supporters praised Floridian government officials for their political and financial support to proceed with the project and the USACE for supervising and managing it.

Guardian Angels and Dutiful Devils: The Devil-Angel Shift

Now that we've laid the foundation of how stakeholders from both sides of the dispute perceive themselves and other stakeholders, we examine how they used the devil-angel shift narrative. The devil-angel shift analyzes how villainously a policy actor portrays other stakeholders (the devil-shift), especially in comparison with how heroically the policy actor identifies themselves (the angel shift) (Shanahan et al., 2018).

From our data, port dredging opponents used the devil-shift 45 percent of the time in their narratives while supporters only used it 5 percent, indicating that narratives aimed at environmental conservation cast opponents as particularly devilish. Surprisingly, the angel-shift was also used more by dredging opponents (22 percent) more than supporters (13 percent). Dredging opponents used the devil-shift to exceedingly villainize the USACE and its contractors in an effort to hold them accountable for the detrimental damages they believed were incurring on the corals and their habitat. There are ample examples in the data. Below are three statements from various members of the Reef Guard, which exemplify how they believed the USACE and its contractors were violating good governance practices:

> *It's outrageous that the Army Corps would stubbornly refuse to learn from its recent mistakes. Florida's coral reefs are a national treasure that deserve to be protected. Rather than diving headlong into the Port Everglades project, the Army Corps needs to step back, learn the lessons of the fiasco at the Port of Miami, and do right by our coral reefs.* (Lopez, 2015)
>
> *The Corps' report misleadingly and conveniently blames the disease and warm temperatures for the destruction of our corals, but all of the available evidence, including evidence from every other federal, state, and local government agency involved, shows that the Corps' dredging that was smothering the reefs long before the disease even began.* (Staletovich, 2019)
>
> *[...] The Army Corps steadfastly refused to incorporate lessons learned in its plans for Port Everglades. Despite claiming a desire for better transparency, the Corps distributed a "Frequently Asked Questions" document that was rife with misinformation [...].* (Silverstein, 2018)

These statements portray the USACE as stubborn, refusing to learn from mistakes, lying to the public, destroying coral reefs, smothering coral reefs and spreading misinformation. Given the stronger likelihood of conservationists using these devil-shift narratives, it is really important that further research explores whether these types of narratives are effective.

Dredging opponents also used the angel-shift to portray themselves as caring stewards aiming for the protection of corals and the surrounding environment and through that lens, justifying their legal challenges. This sentiment is expressed in the following statements made by members of the Reef Guard:

> *We want to give [the Florida Department of Environmental Protection] and [NOAA] and the Corps every chance to get this worked out...But if they're unable to do so, then we're prepared to go forward.* (Staletovich, 2014)
>
> *Our goal is not to stop the dredging but to ensure that the environment is protected and that everyone follows the rules set out when the project was started, and that's not happening.* (Bonney, 2014)

> *The Corps claims it's a 'learning agency,' but all plans so far show that the Corps is not intending to improve its practices in Port Everglades after destroying over 200 acres of reef in Miami, and with this letter we show our intent to push for better protection for Fort Lauderdale's reefs.* (Lopez, 2015)

As a whole, opponents of the port project used the devil-shift tactic frequently, making it evident who the "devils" were (the USACE and its contractors) and who the "angels" were (themselves). Opponents managed to gather significant national-scale attention from prominent newspapers such as the *New York Times*. Furthermore, their narratives influenced the press to agree with them and portray their views favorably in the news, which prominently criticized the USACE and its contractors, one indication that these narrative strategies may be more effective.

Stakeholders that supported the dredging project included government officials holding elected positions, the USACE (which is part of the federal government), some of its contractors and citizens. As such, it is likely that due to their professions, many supporters are more restrained in their narratives and would be frowned upon for severely criticizing non-governmental organizations. This could reflect the low use of the devil-shift tactic in their narrative. However, they did not hesitate to call opponents a group of "obstructionists" using any means to delay the project, as previously noted above. Additionally, while supporters used more subtle language than opponents, their meaning to antagonize was evident, as can be seen in the statement below:

> *We aren't sure what [a leader from Reef Guardians] is saying could have been avoided. He had wanted us to move the pilings earlier in the year to enable large yachts to transit in an unmarked, undesignated channel for the recent Miami boat show [...] Is [this person] implying that he knows of a yacht that killed a manatee? Otherwise, what is he saying that could have been avoided and by what Corps action?"* (USACE Jacksonville District Website, 2014).

Dredging supporters used the angel-shift to strongly praise powerful stakeholders, such as government officials, politicians and port directors. They also commended the progress of the project and its success in balancing the impacts of the dredging project and protecting the surrounding environment:

> *[...] [the] deep dredge project [is] employing some of the highest environmental protection protocols outside designated National Marine Sanctuary [...] USACE with contractors from federal, state, and local agencies [are] working together to ensure the health of Miami harbor is preserved.* (USACE Jacksonville District Website, 2014)

An interview respondent, a coral scientist from a private firm contracted by the USACE, expressed their frustration over the uncompromising and "nasty" language used by dredging opponents to push legal actions (which the respondent believed they did for personal fame and monetary gain) (June 8, 2021). The respondent also remarked that it was impossible to retaliate or respond to any of the defaming comments, which severely tarnished their reputation as a respectable scientist. They further stated that a member from one of the non-governmental organizations opposing the project had a reporter from the *Miami Herald* on speed dial and that the reporting "was biased and definitely not journalism at its finest" (June 8, 2021). It was disheartening to them and to other involved scientists because the respondent believed that they were completing the job as dutifully as was tasked.

Putting aside whether the opponents or supporters were morally justified in their messaging, the devil-shift tactic especially demonstrates how influential non-governmental organizations' narratives can be during environmental disputes over vulnerable ecosystems. Their vilifying narratives convinced the press it was a worthy story to publish and likely helped persuade the court to enhance coral protection through greater coral relocation and restoration efforts funded by the USACE. Our findings also show, at least in their narratives, that stakeholders from both sides of the dispute believed they were acting in good faith. Regardless of true intentions, it is the narratives that are important because they result in policy action, as illustrated in this case study.

The Heated Scientific Dispute in the Port of Miami Case

In the next two sections, we take an in-depth look at how the environmental dispute focused on narratives involving science. Narratives on science concerned the effectiveness of mitigation efforts, the value of the corals that would be destroyed in the shipping channel, and most prominently, the impacts of sedimentation on corals from dredging. Our findings show that dredging opponents used science in 54 percent of their narratives, while dredging supporters talked about science 29 percent of the time.

Dredging opponents were skeptical from the start in 2012 about the mitigation efforts put in place, asking the USACE and other involved actors to limit daily blasting and expand seagrass and coral mitigation projects, arguing that these efforts would not do enough to ensure adequate conservation. Skepticism heightened when contractors discovered "10 times more corals than was found in the previous survey" prior to dredging (Gallardo, 2014). Mistrust reached a peak in 2014, when the public and other government agencies were reporting high rates of sedimentation and extremely cloudy water around the impact site. Throughout the project, opponents sought clarification and condemned the USACE on adverse impacts they believed were associated with the dredging project, as illustrated by quoted statements throughout this section.

Dredging opponents highlighted the lack of transparency and their uncertainty over the methods used to monitor mitigation efforts, expressed in the following statements:

> *While the seagrass project is in clear view, the corals seem to be missing in action [MIA] [...] I'm completely in the dark about where these corals are going [...] Who's checking to make sure they're doing their job [...] Where is this natural reef they've selected? I'd love to know.* (USACE Jacksonville District Website, 2014)

> *[...]State officials noted violations on how work is monitored and offered the Corps assistance [...] One simple solution being worked out will put divers with rulers in the water to measure the sediment rather than rely on complicated devices [...] It's very simple, straightforward and quick to do.* (Staletovich, 2014)

As part of the mitigation efforts, corals were relocated to natural sites or collected to be placed in coral nurseries owned by the University of Miami. Coral scientists permitted to collect corals for the nurseries were surprised to find such a large number of corals in the channel and communicated, about the potential value of the corals as well as a frenzied coral relocation period:

> *These corals may be pre-adapted to deal with warmer temperatures and harsher environments, and it's a huge loss to destroy them when they may be the most resilient corals in the area.* (Udel, 2014)

> *The bad news is that you only have ten days to collect as many corals as possible [...] As scientists, we're used to being so careful [when collecting corals] [...] this was a free-for-all [...] People were going crazy with hammers [...] We had to just collect as many as possible [...] The scientists that work for the government aren't talking about the bigger picture. It's unbelievable.* (Fieseler, 2015)

While opponents criticized the proceedings of the mitigation efforts, they were most concerned about the impacts of sedimentation on the remaining corals. Dredging opponents claimed that sediment plumes extended over 10 miles offshore, that barges were overflowing dropping dredging material on sensitive coral habitat, and that significantly large areas with corals were getting smothered by sediment, leading to extensive coral mortality:

> *The situation is worsened by the fact that the (ocean floor) is now covered with a layer of sediment that will prohibit (future coral) settlement and attachment.* (Fagenson, 2014)

> *The NOAA divers found that sediment depths "were greatest nearest the channel where the dredging was occurring and there was mortality of corals associated with sedimentation that occurred over a fairly broad area.* (Fitzgerald, 2016)

Additionally, an interview with a dive safety officer for quality assurance on the monitoring conducted by private contractors was also alarmed by the high rates of sedimentation and stated, "that there were no actual thresholds in place that would hold or delay construction to minimize or avoid impacts on corals" (June 29, 2021).

Dredging opponents believed that the science (e.g., monitoring and mitigation strategies) was badly conducted and improperly reported throughout the duration of the project. They also felt that the USACE and its contractors "conveniently" blamed the coral bleaching event and disease that emerged during the project for coral mortality, as noted earlier. Finally, opponents expressed certainty in the Corps underestimating impacts from dredging on the corals which led to mass coral mortality:

> *[…] The Corps fails to adequately consider impacts to offshore coral reefs as a result of this project […] These environmental analyses by the Corps in the Draft EA do not rely on the best available science. These planning documents are risky for our coral reefs and seagrass meadows, both of which are disappearing from Biscayne Bay.* (Miami Waterkeeper, 2019)

> *The notion that only six corals died as a result of dredging is actually sort of absurd […] We should really take note of this study to realize that the impacts of dredging on corals can be much more significant than originally planned.* (Staletovich, 2019)

Overall, dredging opponents used science in their narratives to contest the science presented by dredging supporters, to sow mistrust in the governmental and private-sector science being conducted and reported, and to heighten the dispute by drawing attention to potentially severe violations and mishaps. They created the narrative that the USACE and its contractors underestimated the impacts of dredging and sedimentation. This, in turn, caused the USACE to implement insufficient mitigation efforts wherein, bad science was conducted and permits were violated. Dredging opponents also argued that coral mortality was associated with the port project. These failures lent a spotlight on this dispute and caused the Reef Guard to successfully push for legal actions.

In response to these accusations, the USACE and its contractors maintained that they did due diligence, made decisions based on the best science available, and hired the best contractors to successfully implement the mitigation efforts and coral monitoring post-dredging, as demonstrated below:

> *[...] Some of the best scientific agencies and firms in the nation helped develop and carry out the plans for the Miami Harbor deepening project [...] The comprehensive results of a joint study on the effects of deepening the harbor will be available in 2017. The study includes analyzing hundreds of hours of video, photos and documentation from more than 7,000 scientific dives.* (Parker, 2016)

They published news releases on their progress, stating their success in construction and monitoring and relocating coral:

> *Progress is on schedule with the Miami Harbor deepening and widening project, including the successful construction of artificial reefs and relocation of about a thousand healthy corals [...] After 40 days, all of the 38 relocated staghorn colonies were alive and in good health, with only minor bleaching and partial mortality [...] Encouragingly, several colonies were also observed as having new tissue growth over the epoxy base, demonstrating the coral's ability to rapidly adapt to its new environment [...] The mitigation construction and relocations went exceptionally well, and we anticipate a very good survival and growth rate for the relocated corals.* (USACE Jacksonville District Website, 2014)

In response to the discovery of 10 times the amount of coral colonies than reported (31 reported, 243 colonies discovered) prior to dredging, an interview respondent explained that there was miscommunication between the contractors and government organizations (June 8, 2021). The respondent stated that the contractors were following the monitoring procedure given to them and reported the presence of 31 coral colonies accordingly. However, this number should have been multiplied by nine to get the actual number of coral colonies within the estimated dredging impact zone (~279 coral colonies, which would have painted a more accurate picture). The respondent also stated that there was an efficient protocol in place to avoid excessive sedimentation on corals:

> *If there was an emergency, we would notify [the company] Great Lakes [who would] respond within 24 hours. We did this on two to three occasions; where we saw excessive sediment that needed to be addressed. For example, Feb 5, 2014, we sent an email to [Great Lakes who] notified the Corps that day and the Corps came back and told Great Lakes to move the dredge. It was moved by the next night*

> *[showcasing] adaptive management, with a turnaround time of approximately three days.* (June 8, 2021)

Pertaining to the overflowing barges, the USACE maintained that these were also false accusations and that:

> *[...] They are in fact monitoring the corals and the ships are not leaking [...] [they] have special sensors and those detect a draft change [...] And so [they] are monitoring to make sure that no sediment is leaking out of them.* (Gallardo, 2014)

The USACE also refuted the claims that sedimentation was causing mass coral mortality and that any adverse impacts that occurred were expected:

> *Dramatic statements were made about the effects of the dredging in Miami, specifically regarding sedimentation impacts to offshore resources. Contrary to these statements, sedimentation effects were anticipated and are expected to be of short-term duration with no long lasting effect on the ecosystem.* (USACE Jacksonville District Website, 2014)

A major point of contention concerned the duration of sedimentation impact on the corals. While the Reef Guard asserted that sedimentation caused prolonged coral burial leading to coral mortality, the USACE and its contractors argued that coral burial was temporary. Another interview respondent, a scientific diver who logged over 2,000 dives with one of the contracted companies, also observed temporary sedimentation impact:

> *Honestly, I was shocked by how much sedimentation ended up on the reef, definitely saw that change but I was also shocked at how resilient corals were. There would be 30% sediment accumulation one day and the next day almost all sediment was off the coral. I'd never witnessed something like that before. Various oceanographic properties can contribute to that, such as bottom currents.* (June 29, 2021)

The contracted firm responsible for monitoring the tagged coral colonies highlighted that they visited the monitoring sites regularly throughout the project and observed temporal sedimentation rates, as expressed above. They emphasized that other scientists reporting high rates of sedimentation only surveyed the sites a handful of times, thus not seeing the ebb and flow of sediment accumulation and wash-off on corals as they were.

However, the USACE's defenses against Reef Guard's allegations were not strong enough to be upheld in court. In 2014, the USACE settled, funding $400,000 worth of relocation of staghorn coral colonies to a coral nursery supervised by the University of Miami. When the legal case was reopened in 2016, the USACE agreed to secure additional funding to restore 10,000 corals. In the press, the USACE conceded that improvements could have been made regarding the monitoring during the construction and that lessons were learned:

> *We are reporting all of these things [...] We're a learning organization [...] We take the lessons learned and apply them, not only to projects under execution, but to our future planning for projects.* (Alvarez, 2015)
>
> *Our dredging methods and how we monitor that during construction should be improved [...] That's one of the things that was a real lesson learned from Miami.* (Upton, 2016)

Overall, dredging supporters (e.g., the USACE and its contractors) used science in their narratives to assure the public that they implemented sufficient mitigation efforts, which were conducted using the best people and science available. As such, the 2015 Port of Miami dredging project was extremely successful in balancing economic growth and environmental impact. While they acknowledged adverse impacts would occur, it was within the scope of their expectations, and impacts were temporary. From their perspective, the downfall of the project were the false allegations made by dredging opponents (which included the USACE and other government agencies). Likely due to the USACE's position as a powerful government actor, they believed their hands were tied and that it was in their best interest to settle the legal challenges, though they disagreed with the allegations made.

Throughout this chapter but especially in this section, we present two strikingly different narratives of the events that transpired during the 2015 Port of Miami dredging project. We observe how science became the center of this dispute, which resulted in mistrust, tarnished reputations, and litigation. However, our understanding of what actually happened is still far from settled. A similar conflict may emerge for the Port Everglades dredging project, which is still delayed (likely because of the controversial outcome of this project and the COVID-19 pandemic). As of March 2022, there is still an ongoing investigation, and private organizations from both sides of the dispute are arguing in the scientific literature.

The Unsettled Science of Dredging Impacts in the Literature

Impacts of dredging on coral reefs are still debated in the literature. For instance, beginning in 1978, Bak found that the effects of dredging a channel on a fringing coral reef in the Caribbean reduced light access to corals from 30 percent to 1 percent; decreased calcification of coral species by 33 percent;[14] and was fatal to species that were insufficient at rejecting sediment.[15] Similar findings have been observed by the expansion of the Veracruz Port in Mexico. Port construction in the nineteenth and early twentieth century led to 50 percent of direct loss of one reef adjacent to the port and the complete sediment burial of another (Valadez-Rocha & Ortiz-Lozano, 2013). In Brazil, the construction of the Port Complex of Suape altered tidal cycles which negatively impacted the micro-benthic communities on reefs (Costa et al., 2014). Walker et al. (2012) documented shipping-related and port development impacts of the Port of Miami, Port Everglades, and Port of Palm Beach on coral reef habitat. Using topography data beginning in the 1920s, authors discovered that over 300 hectares of reef habitat were impacted and that burial from dredge material was the greatest impact.

However, some researchers have disputed the extent of impacts and sediment burial from dredging close to coral reefs. This contending research states that (1) short-term versus long-term effects of sediment burial should be acknowledged (Brown et al., 1990; Precht et al., 2019), (2) dredging may only have localized impacts (Adjeroud et al., 2016; Precht et al., 2019), (3) corals may recover after dredging (Brown et al., 1990; Precht et al., 2019), and (4) coral mortality may have been due to something other than dredging, such as disease or coral bleaching (Precht et al., 2019). The relationship between increased likelihood of coral disease and presence of dredging has been studied in Australia. One study found that corals exposed to the longest period of sedimentation were twice as likely to be infected with disease, linking dredging with elevated disease presence (Pollock et al., 2014). However, another study refuted this claim and suggested that specific impacts of dredging are difficult to isolate and dredging was not an early indicator of coral disease (Stoddart et al., 2019). These differences suggest the science has yet to be settled and similar disputes will emerge in the future.

Directly linked to the 2015 Port of Miami dredging project, there have been several competing versions and publications concerning the extent of ecological reef damages among scientists. In 2015, Dial Cordy and Associates, the contractors responsible for post-dredging monitoring reported that 219 of

14 Light access to corals is important because they have a symbiotic relationship with a dinoflagellate algae, zooxanthellae, that lives inside coral polyps and photosynthesizes providing the coral with most of of its energy needs. Calcification is an indication of growth rates for hard sclerotin corals, defined as corals with calcium-carbonate skeletons, because they deposit calcium carbonate to grow their skeletal structure.

15 Too much sedimentation leads to burial of corals which is fatal.

612 (36 percent) tagged coral colonies died throughout the dredging project (Precht et al., 2019). Only seven of those tagged coral colonies died due to sedimentation from dredging, while 194 died from, at the time, an unknown white plague disease. The remaining 18 died from coral bleaching, competition and other diseases. As a federal project, this data is in the public domain for anyone to review and use. These findings were called into question by Miller et al. (2016) who found that sedimentation impacts extended 700 meters beyond the dredging area while monitoring was limited within 50 meters of the project. They found some areas had extreme sedimentation accumulation that was likely associated with dredging, suggesting a large number of unaccounted coral deaths. The manager of Dial Cordy and Associates refuted Miller et al.'s (2016) findings by stating that "they did a one-time survey and had no baseline data" (Townshend, 2020).

In 2019, three subsequent papers were published. First, Precht et al. (2019) reviewed and confirmed the findings of Dial Cordy and Associates' 2015 report.[16] Second, Cunning et al.'s (2019) paper found that over 560,000 corals were killed within 500 meters of the dredging site,[17] largely from sedimentation burial and accumulation from dredge-related activities. Using as much of Dial Cordy and Associates' data, Cunning et al.'s (2019) research used statistical analyses to draw trends to examine the full extent of damages incurred instead of focusing on individual data sites. Researchers emphasized that their results were easily reproducible unlike those of Precht et al.'s (2019) (August 20, 2021). Cunning et al.'s (2019) paper that finds over half a million corals died during the dredging project made the headlines of several newspapers, including the *New York Times*. The manager of Dial Cordy and Associates also disputed Cunning et al.'s (2019) findings saying that the scientists cherry-picked the data (Townshend, 2020). Gintert et al.'s (2019) paper used Dial Cordy and Associates' public data as well and found similar results as Precht et al. (2019).[18] Most recently, Spadafore et al. (2021), who also used the public data, supported Cunning et al.'s (2019) finding on partial coral mortality due to sedimentation but more prominently supported Precht et al.'s (2019) findings.[19]

While understanding the impacts of dredging on coral mortality during the 2015 Port of Miami dredging project is important, the relevance of

---

16 The primary author is the director of marine and coastal programs at Dial Cordy and Associates.

17 One of the authors is the CEO of the Biscayne Bay Waterkeeper.

18 Gintert is the second author in Precht et al.'s (2019) paper.

19 According to interview respondent #4, authors in Spadadore et al. (2021) "had no dog in the fight and reanalyzed the data on their own" (June 8, 2021).

the project has shifted to discovering the origins of Stony Coral Tissue Loss Disease. This disease has become one of the most lethal coral diseases, which has impacted over 30 species of corals with a high mortality rate, and has spread rapidly across the Caribbean (Estrada-Salvidar, 2021).[20] The devastating impacts of this disease cannot be overstated.

Stony Coral Tissue Loss Disease directly relates to the 2015 dredging project because the first reported sighting was on September 16, 2014 by Dial Cordy and Associates. In Precht et al. (2019), they state to have observed it during a monitoring survey close to the dredging project. Undoubtedly, this begs the question whether dredging was associated with the emergence of this disease. Multiple interview respondents suggested that this disease was correlated, if not caused, by the dredging activities. Most recently, Studivan et al. (2022) found that reef sediments can transmit Stony Coral Tissue Loss Disease, underlining the need to better understand the role of microbial sediments that can be disrupted during dredging. However, linking dredging to increased likelihood of disease has been disputed in the literature, as outlined above (Pollock et al., 2014; Stoddart et al., 2019).

While Stony Coral Tissue Loss Disease may seem like a recent disease, it wasn't mentioned or on the mainstream radar until 2016, approximately 1.5 years after its discovery (Precht, 2020). Precht (2020) fiercely criticized the slow response of acknowledging this rampant disease. He attributed it to three components. First, he emphasized that the opponents of the project placed a spotlight on dredging impacts on corals instead of focusing on the severity of the disease. Second, he blamed the media for playing into the opponent's hands, and third, he drew attention to the political tension over the acknowledgment of climate change in Florida at the time. Needless to say, Stony Coral Tissue Loss Disease is certainly being talked about now, having swept throughout the Caribbean, leaving a trail of devastating death in its path.

Implications for the Future

In this chapter, we aimed to understand the events that transpired during the controversial 2015 Port of Miami dredging project. We observed the

20 Stony Coral Tissue Loss Disease "been confirmed in the Caribbean off Jamaica, Mexico, Sint Maarten, the US Virgin Islands, Dominican Republic, Turks & Caicos Islands, Saint-Martin, Belize, Sint Eustatius, The Bahamas, Puerto Rico, British Virgin Islands, Cayman Islands, Guadeloupe, St. Lucia, Honduras and Martinique" (AGGRA, 2022).

emergence of an environmental dispute, where one side favors the project while the other contests it. We then examined how stakeholders from both sides of the dispute cast victims, villains and heroes and use the devil-angel shift to push for actions toward their beliefs. Finally, we focused on how the dispute culminated in the interpretation of scientific mitigation efforts. There is still an ongoing dispute, indicating that there remain issues to be resolved.

All sides of the dispute are still left unsatisfied, with everyone stating they were doing what they believed was right. The Reef Guard and opponents believed irreparable damage to the reefs occurred through dredging and later, the emergence of disease, and pushed back through legal actions. And even though they won, "the settlement is bittersweet" because these damages cannot be repaired through greater coral restoration numbers and what is lost, is lost (Staletovich, 2019). An interview respondent despondently said that it didn't help to win the lawsuit, the damage still occurred and questioned what else could have been done (June 4, 2021).

On the other hand, the USACE, its contractors and supporters believed that dredging the port allowed the United States to stay competitive in world trade and increased economic opportunities. They argued that if one of America's largest ports cannot accept the new standard for cargo ships, we will be out-competed. Not only that but they also believed that ample mitigation measures were put in place, ensuring the health of the surrounding environment, including vulnerable seagrass and coral habitat, would not be compromised. So what happens next and what can be learned for the future?

Many recommendations have already been made by Miller et al. (2016), Cunning et al. (2019) and Precht et al. (2019). Miller et al. (2016) point to technical improvements for monitoring, such as collecting live data on sedimentation rates via remote sensing. Cunning et al. (2019) recommend improved dredging policies such as reducing turbidity allowance via regulations and ensuring dredging is offset with coral spawning and times of the year associated with thermal stress. Precht et al. (2019) provide a suite of recommendations, including improved adaptive management capacity, unbiased panel of experts coming toward consensus when an issue arises, increased data collection and monitoring, and liaison, communication and cooperation among all agencies involved. Above all, we find that communication, participation and transparency among all interested stakeholders, the public, agencies and officials are necessary to avoid another environmental dispute as contentious as this one. If not, private actors such as non-governmental organizations can become powerful influencers as observed in this case study. As non-governmental organizations can set their own agenda (and are responsible to their donors), their actions can result in better or worse outcomes. In this case, they directed a spotlight onto potential violations occurring and sought for greater environmental protection. However, Precht (2020) would also argue that they partly impeded the needed management response to stony coral tissue loss disease.

As corals continue to face extensive threats, it is important to understand how dredging projects such as this one impacts them. A number of uncertainties emerged from this project. Did dredging cause mass coral mortality? Is it linked to increased disease susceptibility? Or can coastal development projects proceed while protecting the environment? It is still unclear. However, what is evident is that participatory management to involve and communicate to all interested stakeholders should be better implemented.

References

Adjeroud, M., Gilbert, A., Facon, M., Foglia, M., Moreton, B., & Heintz, T. (2016). Localised and limited impact of a dredging operation on coral cover in the northwestern lagoon of New Caledonia. *Marine Pollution Bulletin*, *105*(1), 208–214. https://doi.org/10.1016/j.marpolbul.2016.02.028

Alvarez, L. (2015, March 7). Despite protections, Miami Port Project Smothers coral reef in silt. *The New York Times*. Retrieved September 14, 2022, from https://www.nytimes.com/2015/03/08/us/despite-protections-miami-port-project-smothers-coral-reef-in-silt.html?_r=1&referrer

Bak, R. P. M. (1978). Lethal and sublethal effects of dredging on reef corals. *Marine Pollution Bulletin*, *9*(1), 14–16. https://doi.org/10.1016/0025-326X(78)90275-8

Biscayne Bay Environmental Enhancement Trust Fund. (n.d.). Retrieved January 30, 2022, from http://miamidade.elaws.us/code/coor_ch24_arti_div4_sec24-40

Bonny, J. (2014, September 5). Environmentalists renew criticism of Miami dredging. *JOC.com*. Retrieved September 14, 2022, from https://www.joc.com/port-news/us-ports/port-miami/environmentalists-renew-criticism-miami-dredging_20140905.html?destination=node%2F2760091

Brown, B., Le Tissier, M., Scoffin, T., & Tudhope, A. (1990). Evaluation of the environmental impact of dredging on intertidal coral reefs at Ko Phuket, Thailand, using ecological and physiological parameters. *Marine Ecology Progress Series*, *65*(3), 273–281. https://doi.org/10.3354/meps065273

Coral Disease Outbreak. AGRRA. (2022, August 2). Retrieved September 14, 2022, from https://www.agrra.org/coral-disease-outbreak/#:~:text=Stony%20coral%20tissue%20loss%20disease%20(SCTLD)%20is%20a%20new%20lethal,quickly%20causing%20high%20coral%20mortality

Costa, D. L., Santos, A. M., Da Silva, A. F., Padilha, R. M., Nogueira, V. O., Wanderlei, E. B., Bélanger, D., Gomes, P. B., & Pérez, C. D. (2014). Biological impacts of the port complex of suape on benthic reef communities (Pernambuco-Brazil). *Journal of Coastal Research*, *30*(2), 362–370. https://doi.org/10.2112/jcoastres-d-13-00055.1

Cunning, R., Silverstein, R. N., Barnes, B. B., & Baker, A. C. (2019). Extensive coral mortality and critical habitat loss following dredging and their association with remotely-sensed sediment plumes. *Marine Pollution Bulletin*, *145*, 185–199. https://doi.org/10.1016/j.marpolbul.2019.05.027

Estrada-Saldívar, N., Quiroga-García, B. A., Pérez-Cervantes, E., Rivera-Garibay, O. O., & Alvarez-Filip, L. (2021). Effects of the stony coral tissue loss disease outbreak on coral communities and the benthic composition of Cozumel reefs. *Frontiers in Marine Science, 8*. https://doi.org/10.3389/fmars.2021.632777

Fagenson, Z. (2014, August 28). Environmentalists ready to sue over Miami Port's deep dredge. *Reuters*. Retrieved September 14, 2022, from https://www.reuters.com/article/us-usa-miami-dredging-idUKKBN0GS2Q720140828?edition-redirect=uk

Fieseler, C. (2015, February 27). Florida dredging would cut path across corals in 'crisis'. *National Geographic Society Newsroom*. Retrieved September 14, 2022, from https://blog.nationalgeographic.org/2015/02/27/this-scientist-made-superhero-headlines-why-doesnt-that-happen-more-often/

Fitzgerald, E. (2016, May 23). A Miami port dredging project may have damaged the city's offshore Coral Reef. *The World from PRX*. Retrieved September 14, 2022, from https://www.pri.org/stories/2016-05-23/miami-port-dredging-project-may-have-damaged-citys-offshore-coral-reef

Frazer, R., Robinson, M. J., & Knobloch-Westerwick, S. (2021). Narratives' impacts on attitudes: Do signaling of persuasive intent and fictionality matter? *Communication Studies, 72*(3), 347–365. https://doi.org/10.1080/10510974.2021.1876127

FWS. (n.d.). *Section 7 consultations*. Retrieved April 19, 2022, from https://www.fws.gov/service/section-7-consultations

Gallardo, C. (2014, August 5). Environmental Group threatens to Sue Army Corps of Engineers about dredging near corals. *WLRN*. Retrieved September 14, 2022, from https://www.wlrn.org/news/2014-07-21/environmental-group-threatens-to-sue-army-corps-of-engineers-about-dredging-near-corals#stream/0

Gintert, B. E., Precht, W. F., Fura, R., Rogers, K., Rice, M., Precht, L. L., D'Alessandro, M., Croop, J., Vilmar, C., & Robbart, M. L. (2019). Regional coral disease outbreak overwhelms impacts from a local dredge project. *Environmental Monitoring and Assessment, 191*(10), 1–39. https://doi.org/10.1007/s10661-019-7767-7/figures/8

Gloor, C. (2010, November 23). Port of Miami project serves as economic growth for South Florida. *Miami's Community News*. Retrieved September 14, 2022, from https://communitynewspapers.com/miami-beach-news/port-of-miami-project-serves-as-economic-growth-for-south-florida/

Lopez, J. (2015, September 4). Widespread coral damage at Port of Miami project sparks call to Reevaluate Port Everglades expansion. *Center for Biological Diversity*. Retrieved September 14, 2022, from https://www.biologicaldiversity.org/news/press_releases/2015/corals-09-04-2015.html

McGrory, K. & Morgan, C. (2012, April 25) Port of Miami project gets help from Tallahassee - State lawmakers want to speed up the Port of Miami Deep Dredge project, which is being held up by legal challenges. *Miami Herald*. https://infoweb-newsbank-com.spot.lib.auburn.edu/apps/news/document-view?p=AWNB&docref=news/13D1CA82E8B24238&f=basic

Miami Waterkeeper (2019, April 25). Army Corps Finds No Significant Impact In Port Miami Channel Maintenance Dredging. We Disagree. https://www.miamiwaterkeeper.org/army_corps_finds_no_significant_impact_in_portmiami_channel_maintenance_dredging_we_disagree

Miller, M. (2012, April 25). Port of Miami deep dredge clears final hurdle as environmentalists drop lawsuit. https://www.miaminewtimes.com/news/port-of-miami-deep-dredge-clears-final-hurdle-as-environmentalists-drop-lawsuit-6519362

Miller, M. W., Karazsia, J., Groves, C. E., Griffin, S., Moore, T., Wilber, P., & Gregg, K. (2016). Detecting sedimentation impacts to coral reefs resulting from dredging the Port of Miami, Florida USA. https://doi.org/10.7717/peerj.2711

Muller, E. M., Sartor, C., Alcaraz, N. I., & van Woesik, R. (2020). Spatial Epidemiology of the Stony-Coral-Tissue-Loss Disease in Florida. *Frontiers in Marine Science*, *7*, 163. https://doi.org/10.3389/FMARS.2020.00163/BIBTEX

NOAA Fisheries. (n.d.b.). Permits for the incidental taking of endangered and threatened species. Retrieved April 19, 2022, from https://www.fisheries.noaa.gov/permit/permits-incidental-taking-endangered-and- threatened-species

Parker, S. (2016, December 13). Miami harbor dredging claims not moot, Enviros Say. *Law360*. Retrieved September 14, 2022, from https://www.law360.com/articles/871812/miami-harbor-dredging-claims-not-moot-enviros-say

Pollock, F. J., Lamb, J. B., Field, S. N., Heron, S. F., Schaffelke, B., Shedrawi, G., Bourne, D. G., & Willis, B. L. (2014). Sediment and turbidity associated with offshore dredging increase coral disease prevalence on nearby reefs. *PLoS ONE*, *9*(7), 102498. https://doi.org/10.1371/journal.pone.0102498

Precht, W. F., Aronson, R. B., Gardner, T. A., Gill, J. A., Hawkins, J. P., Hernández-Delgado, E. A., Jaap, W. C., McClanahan, T. R., McField, M. D., Murdoch, T. J. T., Nugues, M. M., Roberts, C. M., Schelten, C. K., Watkinson, A. R., & Côté, I. M. (2020). Chapter Twelve - The timing and causality of ecological shifts on Caribbean reefs. In B. M. Riegl (Ed.), *Population dynamics of the reef crisis* (Vol. 87, No. 1, pp. 331–360). Academic Press. https://doi.org/10.1016/bs.amb.2020.08.008

Precht, W. F., Gintert, B. E., Fura, R., Robbart, M. L., Rogers, K., & Dial, R. S. (2019). Miami harbor deep dredge project: A reappraisal reveals same results. *Dredging Summit and Expo'19 Proceedings*, June, 4–31. www.westerndredging.org

Rodrigues, D. C., Vasconcellos, M., & Vasconcellos, A. M. de A. (2020). Advocacy coalition formation and key actors within the policy. *Revista de Administração Pública*, *54*(6), 1711–1728. https://doi.org/10.1590/0034-761220190450X

Shanahan, E. A., Jones, M. D., Mcbeth, M. K., & Radaelli, C. M. (2018). The narrative policy framework. In *Theories of the policy process* (pp. 173–213). https://doi.org/10.4324/9780429494284-6

Silverstein, R. (2018, June 5). Army corps' 'alternative facts' bad news for our reefs: Opinion. *Sun Sentinel*. Retrieved September 14, 2022, from https://www.sun-sentinel.com/opinion/commentary/fl-rsoped-port-everglades-expansion-20170327-story.html

Spadafore, R., Fura, R., Precht, W. F., & Vollmer, S. V. (2021). Multi-variate analyses of coral mortality from the 2014–2015 stony coral tissue loss disease outbreak off Miami-Dade County, Florida. *Frontiers in Marine Science*, *8*, 1299. https://doi.org/10.3389/fmars.2021.723998/bibtex

Staletovich, J. (2014, September 17). Feds say 'nurseries' could save government cut Coral. *Miami Herald*. Retrieved September 15, 2022, from https://www.miamiherald.com/news/local/environment/article2144896.html

Staletovich, J. (2016, February 1). Corps says disease, not dredging, hurt PortMiami Coral. *Miami Herald*. Retrieved September 15, 2022, from https://www.miamiherald.com/news/local/environment/article57756768.html

Staletovich, J. (2019). Latest chapter. In *Port Miami Dredge: More dead coral, biologist charged with lying*. WLRN. https://www.wlrn.org/environment/2019-06-05/latest-chapter-in-port-miami-dredge-more-dead-coral-biologist-charged-with-lying

Stoddart, J., Jones, R., Page, C., Marnane, M., De Lestang, P., & Elsdon, T. (2019). No effect of dredging on the prevalence of coral disease detected during a large dredging program. *Marine Pollution Bulletin, 140*, 353–363. https://doi.org/10.1016/j.marpolbul.2019.01.047

Studivan, M. S., Rossin, A. M., Rubin, E., Soderberg, N., Holstein, D. M., & Enochs, I. C. (2022). Reef sediments can act as a stony coral tissue loss disease vector. *Frontiers in Marine Science, 8*. https://doi.org/10.3389/fmars.2021.815698

Towle, E. (2020). A status report for Florida's Coral Reef 2020. https://www.coris.noaa.gov/monitoring/status_report/docs/FL_508_compliant.pdf

Townshend, A. (2020, September 8). Firm responsible for Miami Harbor Deepening project coral monitoring stands by its data and results. *The Waterways Journal*. Retrieved September 14, 2022, from https://www.waterwaysjournal.net/2019/11/06/firm-responsible-for-miami-harbor-deepening-project-coral-monitoring-stands-by-its-data-and-results/

Udel, D. (2014, June 4). University of Miami scientists mobilize to save corals from deep dredge in Miami's Port Entrance. *University of Miami News and Events*. Retrieved September 14, 2022, from https://news.miami.edu/rsmas/stories/2014/06/university-of-miami-scientists-mobilize-to-save-corals-from-deep-dredge-in-miamis-port-entrance.html

Upton, J. (2016, May 9). Florida dredging would cut path across corals in 'crisis'. *Florida Dredging would cut path across corals in 'crisis' | Climate Central*. Retrieved September 14, 2022, from https://www.climatecentral.org/news/florida-dredging-corals-in-crisis-20333

US Department of Commerce. (n.d.). *What is dredging?* NOAA.

USACE-Jacksonville District Website. (2014). Retrieved September 14, 2022, from https://www.saj.usace.army.mil/Media/Beyond-the-Headlines/Beyond-the-Headlines-Port-of-Miami/

Victory alert: PortMiami legal settlement. Miami Waterkeeper. (2018, August 6). Retrieved September 14, 2022, from https://www.miamiwaterkeeper.org/victory_alert_portmiami_legal_settlement

Valadez-Rocha, V., & Ortiz-Lozano, L. (2013). Spatial and temporal effects of port facilities expansion on the surface area of shallow coral reefs. *Environmental Management, 52*(1), 250–260. https://doi.org/10.1007/s00267-013-0098-5

Wagner, P., & Ylä-Anttila, T. (2018). Who got their way? Advocacy coalitions and the Irish climate change law. *Environmental Politics*. https://doi.org/10.1080/09644016.2018.1458406

Walker, B. K., Gilliam, D. S., Dodge, R. E., & Walczak, J. (2012). Dredging and shipping impacts on Southeast Florida coral reefs. *Marine & Environmental Sciences Faculty Proceedings, Presentations, Speeches, Lectures. 45*. https://nsuworks.nova.edu/occ_facpresentations/45

Chapter 7

HOW COMMUNITIES ARE ORGANIZING TO CONTEST MAJOR INFRASTRUCTURE PROJECTS THAT DAMAGE CORAL REEFS: THE CAYMAN ISLANDS CASE

By Sabine Bailey and Kelly Dunning

You're about to embark on your long-awaited Caribbean cruise vacation, beginning at the Port of Miami, destined to stop by the elegant and exclusive Cayman Islands. After cruising for two days, you spot Grand Cayman Island, surrounded by turquoise blue waters, lush vegetation and colorful buildings. The ship stops just before the island and anchors its position. A smaller boat comes to greet you at the bottom of the cruise ship's stairs, where you are taken onboard the small vessel. With the wind whipping in your hair and the ocean at your fingertips, you travel less than ten minutes before the small boat docks at the George Town Cruise Port. You disembark and are excited to begin your visit to Grand Cayman and indulge in the so-called "CaymanKind" experience, or the well-known warmth of the Caymanian people, and all that it has to offer.[1] You hop into a taxi and begin your adventures swimming with stingrays at Stingray City, exploring crystal caves and relaxing at Seven Mile Beach. You return to George Town Harbour earlier than expected and observe visitors and locals alike snorkeling close to the

1 CaymanKind originated in 2011 from an advertisement campaign which represented Cayman's hospitality and welcoming nature to visitors. It has become a cultural expression in the Cayman Islands symbolizing the guaranteed kindness and unique experience visitors will have visiting the islands (*Cayman Islands Marketing | Caymankind: Advertising Campaign*. n.d.).

port. You rent a mask, snorkel and fins, and directly from the port, snorkel the reefs surrounding the harbor, which are teeming with fish and pristine coral reefs. This concludes your visit to the Cayman Islands and you get in line to board the small boat to return to the cruise ship. Just before you embark on the small vessel, you snap an idyllic picture of the towering cruise ship from afar, resting neatly atop the crystal blue waters.

The Cayman Islands and Its Harbor

This would be a general experience that many cruise passengers would have visiting the Cayman Islands. The Cayman Islands consist of three Caribbean islands: Grand Cayman, Little Cayman and Cayman Brac. They lie directly south of Florida, approximately 450 miles away. They are a self-governing United Kingdom overseas territory. As such, they must comply with some international commitments and regulations as well as their own. The Cayman Islands have a population of just over 67,000 people, most of whom reside on Grand Cayman. Finance and tourism are the two economic pillars of the Cayman Islands.[2,3] Tourism revenue comes from overnight visitors, called "stay-over" visitors, and cruise passengers (who tend to be day-trippers who spend the day on the island and leave before nightfall) and contributes between 20 and 25 percent of the nation's gross domestic product (Matthews, 2020). Stay-over tourists only make up 20 percent of tourists but contribute over 70 percent of the economic revenue from the tourism sector in the Cayman Islands (Cayman News, 2019; López, 2021). Prior to the COVID-19 global pandemic, in 2019, the Cayman Islands saw a total of ~2.33 million tourists, 1.83 million from day-tripper visitors and over 500,000 from stay-over visitors (López, 2021). While cruise tourism contributes 30 percent of tourism revenue, it is important because the cruise industry makes up the remaining 80 percent of Cayman's visiting tourists and still contributes over $200 million annually. What all of this means is that cruise tourism is an essential industry within the Caymanian economy.

2 As an international financial center with a tax-neutral platform, large multinational corporations, hedge funds and others are attracted to the financial services provided by the islands, such as offshore banking services and fiduciary services (Boxall, 2021; Cayman Islands Government, n.d.). The finance sector is vital to the economy, contributing 44 percent to the Cayman Islands' gross domestic product (Ministry of Financial Services, 2021).

3 The Cayman Islands do not impose taxes, including corporate, income, property, capital gain, payroll, withholding taxes making them a tax-neutral nation.

Cruise passengers enter the Cayman Islands from the George Town Cruise Port, which is located in George Town Harbour on the western end of Grand Cayman (see Figure 7.1). The George Town Cruise Port is called a "tender port" because large cruise ships cannot dock directly onto the port because it is too shallow for their deep hulls to dock safely. Instead, cruise ships anchor offshore and small boats, called tender boats, transport passengers safely to the island from the ship and vice versa. This system has been criticized as one that could negatively impact the visitors' experience and result in a greater number of canceled tours when the weather is too extreme for tender boats to travel safely. Cruise lines such as Royal Caribbean and Carnival Cruise have publicly announced that they are likely to remove locations that cannot

Figure 7.1 *Top:* Map of Grand Cayman island. *Bottom:* The George Town Cruise Port and locations of adjacent reefs and wrecks threatened by the new cruise berthing facility project. Adapted from Google Earth.

accommodate cruise ships to directly dock onto port from their itineraries (Kirkconnell, 2018; McCarthy, 2018).[4] As such, the former Cayman Islands Government stated that they would build a new cruise berthing facility to upgrade the George Town Cruise Port to allow cruise ships to dock directly onto the port. Unfortunately, this would directly impact and likely destroy 15 acres (approximately 11 American football fields) of pristine coral reefs that surround the George Town Cruise Port. Given the enormous environmental impact and cultural importance of these reefs, a major environmental dispute emerged, remnants of which are still present as of March 2022.

This chapter first explores how the environmental disputes over the new cruise berthing facility project unfolded. We then examine how coalitions, those supportive and those against the project, influenced policy change throughout the dispute. Specifically, drawing from the public policy literature, we look at three different pathways toward policy change driven by coalitions: external events, internal events and policy-oriented learning (Jenkins-Smith et al., 2018). We describe each of these pathways in further detail later on.

The following sections present our case study data with methods and data collection information contained in the Appendix for Chapter 7.

The Dispute of the New Cruise Berthing Facility Project

In 2013, the former Cayman Islands Government made the announcement that they would proceed with a new cruise docking or "berthing" facility in the George Town Harbour that included two extended piers, land reclamation and dredging works (see Figure 7.2).

However, George Town Harbour is surrounded by several popular and vibrant reefs that attract tourists from all over the world. These include Balboa's Reef and Wreck, Eden Rock, Soto's Reef and Devil's Grotto. These reefs hold cultural, historic, ecological and economic significance to the Caymanian people. For example, Soto's Reef is named after Bob Soto, who is widely credited for establishing the first dive shop in the world in 1957 as well as the overall recreational diving industry in the Caribbean (Dunning, 2021). The historic Balboa shipwreck, which sank in a hurricane in the 1930s, also holds cultural heritage importance. Furthermore, each acre of reef in the George Town Harbour is worth approximately $1,630,860 annually from tourism, coastal protection and habitat for marine organisms, values accrued

4 Carnival and Royal Caribbean are two of the world's largest cruise corporations and make up the majority of cruise passenger arrivals.

Figure 7.2 *Left:* Current state and layout of the George Town Cruise Port area. *Right:* Future model of the George Town Cruise Port for the proposed new cruise berthing facility project. Adapted and taken from *supportourtourism.com.*

to humans known as ecosystem services (Baird, 2015). The reefs are also unique due to their ease of access, where hundreds of visitors and Caymanians can enjoy these reefs with a quick swim from a public parking lot onshore, meaning people do not need to own a boat to access vibrant coral reef ecosystems (Dunning, 2021). According to an environmental impact assessment conducted by a private-sector firm Baird in June 2015,[5] the project footprint would directly impact ~15 acres of coral reef and associated marine habitat, which included the aforementioned reefs and wrecks of Caymanian cultural and ecological importance (Baird, 2015). Coral mitigation efforts, such as coral relocation and restoration were proposed to reduce adverse coral reef impacts.

Governance of coral reefs in the Cayman Islands

The Cayman Islands have strong international and domestic policies to protect their reefs (see table A.7.1). In 1992, the United Kingdom signed the Convention on Biological Diversity,[6] which is a legally binding international treaty that tasks signatories to develop and implement conservation management plans to preserve species and ecologically important habitat. Upon the request of the Cayman Islands Government, this commitment and with it, the adoption of the treaty's Aichi Targets was extended to the Cayman Islands. The Aichi Targets aim to reduce direct pressure on biodiversity, address underlying causes of biodiversity loss across signatory governments,

5 Baird is a privately owned environmental and engineering consulting agency that was contracted by the former Cayman Islands Government.

6 The Cayman Islands is a UK Overseas Territories (UKOT) which is under the formal control and sovereignty of the UK but is not part of the UK and is self-governing.

preserve ecosystems and enhance ecosystem services and biodiversity benefits for all (*Aichi Biodiversity Targets*, 2020; Department of Environment, 2009). Furthermore, under Principle 4 of the Cayman Islands Environment Charter,[7] a legally binding document with the United Kingdom government, the Cayman Islands Government is required to perform environmental impact assessments and accept stakeholder involvement in the decision-making process across government agencies, environment and heritage organizations, and the public for major development projects that substantially impact the environment (Cayman Islands Environment Charter, 2001).

Domestic policies that protect reefs in the Cayman Islands include the long-standing Marine Conservation Law. The law was first implemented in 1978, which helped establish MPAs in the Cayman Islands and penalized the unlicensed taking of all coral species (*The Marine Conservation Law, 1978 (1995 Revision)*, 1995).[8] Other regulations include Part I of the Cayman Constitution (2009), which orders the government to limit ecological degradation, advance conservation and ensure that economic or social developments are ecologically sustainable for present and future generations (*The Cayman Islands Constitution Order*, 2009). Prompted by the Convention on Biological Diversity treaty, the Cayman Islands developed a National Biodiversity Action Plan (2009), Habitat Action Plan (2009), and Species Action Plan (2009) for the Cayman Islands with a simple goal to achieve "zero extinction in the Cayman Islands" (Department of Environment, 2009). More specifically, the coral reef Habitat Action Plan (2009) works toward placing all coral reefs under legislative protection, recovering degraded reefs, establishing new marine protected areas that incorporate 50 percent of coral reef habitat and opposing developments that threaten net loss or damage coral reefs (Department of Environment, 2009). Finally, the National Conservation Law (2013) is a strong piece of legislation that protects all coral species at all times and prohibits any development that adversely affects a protected area or priority species (see Table A.7.1 in Appendix 7 for an overview of the Cayman Islands policies).

The Cayman Islands Environmental Assessment Board reviewed the 2015 environmental impact assessment. They expressed that the environmental impacts, especially on coral reefs, were likely underestimated. Furthermore, the success of mitigation efforts, including wreck and coral restoration and

7 Environment Charter Commitments between the United Kingdom and UKOTs ensure shared responsibility of the conservation of the UKOT's environment and international commitments.

8 Part of George Town Harbour used to be inside a marine reserve, however it was recently converted to a port anchorage area which is presently being legally disputed.

relocation were overestimated (Dunning, 2021).[9] The government also solicited public comment pertaining to the findings of the environmental impact assessment, many of which did not support the project. Despite the stringent environmental policies in place, the environmental impact assessment review, and negative public comments, the former Cayman Islands Government announced that they would proceed with the cruise berthing project four months later, in October 2015.

Beginnings of a referendum

The Environmental Assessment Board review and announcement raised concerns among some members of the public. Citizens began a campaign to collect signatures of 25 percent of eligible Caymanian voters to trigger a people's initiated referendum,[10] which would be a first of its kind for all UK Overseas Territories. The referendum was an opportunity to allow citizens to vote on whether they wanted the proposed cruise berthing facility project to proceed. Campaigners formed a coalition called *Cruise Port Referendum* (CPR Cayman) which was led by interest groups, the National Trust of the Cayman Islands,[11] and Opposition members of the legislative assembly. In June 2019, CPR Cayman successfully collected over 5,000 signatures and handed them over to the Cayman Islands Government for verification.

In July 2019, the former Cayman Islands Government announced private-sector actor Verdant Isle Port Partners as the preferred bidder to design, finance, build and manage the ~$240 million dollar berthing facility project.[12] The government claimed that this financial model would pose no financial risk to the Caymanian people and it would not involve any govern-

9 The Environmental Assessment Board, a subset of the National Conservation Council, includes the Port Authority, National Roads Authority, the Department of Planning, Department of the Environment, and the National Museum (Dunning, 2021).

10 Under Section 70 of the Cayman Constitution (2009), a people's-initiated referendum can be triggered via petition.

11 The National Trust is a non-governmental organization with statutory authority established in 1987 with the purpose of preserving natural resources and historical sites under Cayman Law.

12 Verdant Isle Port Partners is a consortium which consists of four businesses including McAlpine Ltd, Carnival Corporation, Royal Caribbean Cruises Ltd. and Orion Marine Construction Inc. Carnival and Royal Caribbean are two of the world's largest cruise corporations, McAlpine Limited is a construction company in the Cayman Islands, and Orion Marine Construction Inc. is an international marine construction company.

Figure 7.3 New project design offered by the Cayman Islands Government and Verdant Isle Port Partners in an effort to minimize impacts on surrounding reefs. Taken from supportourtourism.com.

ment loans or bonds However, the government would surrender $2.32 of the head tax, defined as a uniform tax for each cruise passenger entering the Cayman Islands, to VIPP over the next 25 years (Klein, 2019).

In 2018, stakeholder networks supporting the project also formed a coalition called *Support Our Tourism* in response to the opposing coalition's campaign. It was led by elites within the regime, mainly the premier and their cabinet, the Ministry of Tourism, and later on, private-sector leadership from Verdant Isle Port Partners, the preferred bidder in the private sector. Once the preferred bidder was announced, Support Our Tourism launched an updated campaign that they claimed was more environmentally friendly, putting forth a new project design that consisted of two extended finger piers to mitigate the damage to coral reefs (see Figure 7.3). However, the environmental impact assessment was not redone to assess the ecological impacts of this new project design. Support Our Tourism also pledged to replant 30–50 percent of the affected coral,[13] relocate 100 percent of threatened coral species and provide 10:1 mitigation efforts, in which they would replant ten corals

13 Replanting a coral means either moving a coral from one location to another by placing it onto substrate for coral to attach to or growing corals in a tank and once large enough, planting the small coral onto substrate for corals to attach to and grow.

for every coral that was lost due to the project. They claimed that the project would *increase* coral cover and biodiversity due to this mitigation plan.

In September 2019, the Cayman Islands Election Office verified the petition and the Cayman Islands Government set a referendum date for December 19, 2019. The procedure on how to set up the referendum was unclear due to lack of precedent. For the vote to stop the project from proceeding, the government stated that 51 percent of eligible voters, as well as Caymanian citizens abroad, would have to vote "no." Additionally, any no-show of eligible voters would be counted as a de facto "yes" vote to the project continuing. The opposing coalition believed that these voting policies were unfair and unconstitutional, especially as many citizens would be abroad because of Christmas, which was only six days away from the set referendum date, all but guaranteeing a yes vote.

As such, a member of CPR Cayman and the conservation organization the Cayman Islands National Trust filed legal challenges in court, which later combined into a single lawsuit, against the Cayman Islands Government to delay the referendum. The legal challenge claimed that (1) the Cayman Islands Government misinterpreted and violated the law on how to proceed with the referendum, (2) the newly proposed design required an updated environmental impact assessment before a vote could be called for, and (3) that the port project violated Cayman laws. The court agreed with the lawsuit, and the referendum was delayed into 2020. However, with the onset of the COVID-19 pandemic in March 2020, the cruise berthing facility project came to a halt, and with it, the urgency to set the referendum.

In July 2020, an appeal made by the government overturned the ruling that the referendum had been unconstitutional. Presently, CPR Cayman is still asking for the referendum to ask whether a future cruise berthing facility project could be undertaken by the government. However, the former government denied their request stating that the referendum was specific to the current project alone. A general election in April 2021 resulted in a new government administration that announced they would not proceed with the project during their term. However, Royal Caribbean has made it clear that they are still undecided on how to proceed and that this project is temporarily halted (Whittaker, 2021). This tells us that, once the pandemic ends and cruise tourism recovers, the same coalitions may emerge to lobby for or against a cruise berthing facility project.

Policy Change Driven by Coalitions

This case study presents the dispute regarding the environmental and economic impact of the cruise berthing facility project among networks of stakeholders that began in the early 2010s. Two distinct networks emerged as leaders on either side of the dispute. Supporters formed a coalition called Support Our Tourism that advocated for economic growth and what they argued would be substantial economic opportunities brought by the cruise berthing project via the cruise industry. On the other hand, opponents formed CPR Cayman to trigger the first people's initiated referendum in Caymanian history to suspend the project, an experiment in democratic processes and legal maneuvering aimed at ensuring laws on the books were applied, and that cruise companies were not allowed to break the laws in the event that significant economic gain was on the table.

In the prior section, we explored how the events of the environmental dispute over the proposed project proceeded. To further analyze these types of environmental disputes, we examine how policy change was driven by the two coalitions. We argue that the mobilization of two coalitions, especially project opponents in CPR Cayman, was integral to major policy change, which is evidenced by the current indefinite halt on the cruise berthing facility project.

Our approach: Pathways to policy change

Policy change can be evaluated by the differences between the current policy and previous policy. Major policy change occurs when the policy subsystem (i.e., government plans for a cruise berthing facility project) undergoes a significant shift in one direction. This could be due to transitions between government administrations, or the termination, halting or launching of major programs and projects. Minor policy change usually involves modifications of the pathways to achieve desired policy outcomes. According to the public policy literature that examines advocacy coalitions, there are four pathways to the process of generating policy change: external events, internal events, negotiated agreement and policy-oriented learning (Jenkins-Smith et al., 2018).

First, external events resulting in disturbances in the policy subsystem are often the event that triggers the changes themselves, but are not sufficient for enactment of major policy change. External events are typically attributed to natural disasters or the mobilization of minority coalitions (e.g., protests). Together with elevated political and public attention, strategic narratives[14]

14 Narratives can play an outsized role in shaping our perceptions and decisions, which can lead to policy change. As such, strategic narratives can be employed by policy actors to push their policy beliefs. For example, policy actors can highlight risks in

and the distribution of resources within the coalition, these elements can disrupt the policy subsystem toward major policy change.

The second pathway to policy change includes internal events that occur within the subsystem, such as policy scandals, failures and crises. Internal events, by contrast with external events, can lead to major policy change. This is because coalitions can exploit these internal events to spread doubts about policy effectiveness of the dominant coalition. The third pathway to major policy change can arise through negotiated agreements between opposing coalitions, which is most likely to occur when all coalitions are unhappy with the status quo and seek change by working together. We don't explore this pathway because there were no instances of negotiated agreements among stakeholders in our case study.

The first three pathways can induce major policy change. The fourth pathway, policy-oriented learning, usually results in minor policy change, which typically occurs gradually over long periods of time. Policy-oriented learning is defined as coalitions and actors revise or shift prior belief systems and intentions as a result of gained experience, exposure, knowledge or for strategic purposes (Jenkins-Smith et al., 2018). While major policy change is possible through policy-oriented learning, it is generally in combination with another pathway (e.g., external events, internal events, negotiated agreements) of policy change. Data is summarized in Appendix 7 in tables A.7.2 and A.7.3)

Three Pathways to Policy Change in the Cayman Islands Port Dispute

External events

External events disrupt the system. In the case of the dispute over the reefs of Georgetown Harbour, the policy subsystem is the network of agencies, policies, businesses and prominent political actors involved with the cruise berthing facility project. Events that disturbed the project's progression began with the online petition that was sent out after the publication of the environmental impact assessment report in 2015. It garnered international attention resulting in large organizations such as the Dive Equipment and Marketing Association (DEMA) and Professional Association of Dive Instructors (PADI) to write to the Cayman Island government to reconsider the project. These groups also encouraged their supporters to sign the petition against the port

their narratives to continue the debate or place responsibilities of bad consequences on others to hold them accountable and to accelerate policy action.

project. The environmental dispute, in an account that was favorable to the coalition opposing the port project, was also highlighted in prominent news platforms, such as the British Broadcasting Company (BBC).

International scientists and consultants from the United States and Australia also joined the environmental dispute offering their views on the value of Cayman's pristine coral reefs. They provided their expertise on the feasibility of coral relocation and restoration, highlighting the uncertainties of these mitigation efforts. They emphasized that the success of coral relocation, especially at the scale that was being promised by Support Our Tourism (10:1 mitigation), had yet to be put into practice and success was undetermined and highly uncertain (Boström-Einarsson et al., 2020; Ware et al., 2020). This helped confirm the opposing coalitions' stance in contesting the project.

The formation of anti-port project coalitions including Save Cayman, and ultimately CPR Cayman, was integral to elevating the debate onto the national level, a prominence which continues to this day. The coordination level between actors within this coalition was high and evident through the multitude of events CPR Cayman organized throughout 2018 and 2019. Outreach efforts included sharing information via social media, TV, press and radio, tabling events, organizing public meetings, running boat tours and organizing public events including a Christmas rally. The most important engagement event was undoubtedly the door-to-door canvassing of this coalition to collect signatures to enact the first-ever people's initiated referendum to stop the port project. These events were all aimed at engaging and educating the public on the impacts that opponents believed the cruise berthing facility would have on the surrounding reefs and the potential economic burden for Caymanians. CPR Cayman coalition members strongly emphasized two points: (1) how the referendum was a historic moment and (2) how the referendum was part of Caymanian citizens' constitutional rights to have their voices heard. The efforts put forth by CPR Cayman members and supporters proved to be successful in triggering the unprecedented referendum by December 2019.

The success of triggering the first people's initiated referendum in Caymanian history resulted in a major policy change in the subsystem. This event forced the Cayman Islands Government to press pause on the project and set up the referendum.

The triggering of the people's initiated referendum via signatures of the electorate also led to policy change outside the subsystem and into the interpretation of Section 70 of the Cayman Island Constitution (2009). This was the result of the legal challenges the government faced to delay the flawed referendum they had set up. These challenges resulted in a ruling that the referendum violated Cayman law. Although this ruling was overturned, the onset of the COVID-19 pandemic in early March 2020 further halted the progression of the project. Finally, general elections in April 2021 resulted in a new government that stated they would not engage with the cruise berthing facility project during their term. Both the impacts of the COVID-19 pandemic and the change in government administration are further evidence of major policy change to the proposed cruise berthing facility project through external events.

Internal events

Internal events such as political scandals, failures and crises can also lead to major policy change. This is because one coalition can exploit internal events to spread doubt about the policy supported by an opposing coalition. This case presents evidence of this occurring, when politicians within the Cayman Islands Government took controversial actions that were leveraged by CPR Cayman to block the port project.

Internal events impacting the two coalitions began with the publication of the environmental impact assessment and a kick-off public meeting in 2015, which outlined that at least 15 acres of pristine coral reefs would be destroyed. This information sparked a crisis, which led to the launch of the online international petition as well as the formation of an interest group called Save Cayman opposing the project. In 2018, information came to light that the Cayman Islands Government prevented several waterfront shops from re-entering into their contracts to continue their businesses, signaling that the government intended to go ahead with the port project despite public opposition and lingering questions about uncertainty over environmental impacts. This galvanized the formation of the interest group CPR Cayman, who used this additional crisis to galvanize opposition to the project.

The crisis intensified during the next public meeting in 2018, where it was revealed that the Department of Environment was absent from this meeting, and from its statutory role in providing oversight. Additionally, the director of the Department of the Environment had been removed from the steering committee, a role carved out by Caymanian regulations. The Cayman Islands Government justified the absence by stating that they no longer had a role in the process for the Department of the Environment as they had already reviewed the environmental impact assessment. This caused an additional crisis where some members of the public and the port project opposition coalition viewed these actions with mistrust. Given that attendees of these public meetings had environmental questions that could not be answered by experts at public meetings aimed at transparency, opponents argued that removing the director of the Department of Environment from the committee was a mistake made by the government. They believed it showed a lack of commitment and transparency to protect the reefs that surrounded George Town Harbour.

Shortly afterward, the port project opposition coalition called upon the government to hold a referendum to allow the Caymanian public vote on whether the project should proceed. Political leaders denied this request. Following this denial, CPR Cayman began holding a series of public meetings to voice their concerns over the project and their opposition. These meetings helped gain momentum for CPR Cayman to collect signatures that would be required to force a first-ever people initiated forward. Ultimately, the coalition opposing the port project would collect enough signatures, which then had to be verified by the government one by one, to hold this referendum, an action which required widespread buy-in.

Once the people-initiated referendum had been verified, and during subsequent public discussions regarding the referendum, the CEO of multinational cruise company Royal Caribbean was invited to the Cayman Islands, which the project's opposition claimed was inappropriate. CPR Cayman saw this as a tactic to intimidate Caymanians, since this CEO mentioned that Cayman could be removed from Caribbean cruise itineraries if they did not build piers. The visit, opponents argued, was to use the power of the multinational company to persuade the public to vote "yes" to proceed with the berthing facility project or risk economic crisis and a loss of cruise tourism. The government denied this claim. Lastly, the handling and setup of the people-initiated referendum was poorly received by the anti-port networks and led to lawsuits which further delayed the referendum and project.

Port project opponents cited the lack of transparency and freely available information from trusted scientists like the staff of the Department of the Environment concerned the public the most. This frustration culminated in the collection of signatures to successfully trigger the referendum. While the government consistently claimed that information could not be released until the preferred bidder was announced (preferred bidder is the company given the government contract to build the port infrastructure project). But, by that time, the CPR Cayman campaign was already well underway and only gaining momentum toward successfully triggering the first people-initiated referendum in Caymanian history.

Policy-oriented learning

Policy-oriented learning occurs when coalitions and political actors revise or shift prior statements and intentions. That can be due to gained experience, exposure, knowledge or for strategic purposes (Jenkins-Smith et al., 2018).

These shifts usually result in minor policy changes; however, major policy change is possible, usually in combination with external or internal events. Policy-oriented learning is also most likely facilitated when there are opportunities for all stakeholders to be involved, such as public meetings.

We found instances of policy-oriented learning from both the supporting and opposing coalitions. The first instance occurred shortly after the public meeting on the environmental impact assessment in 2015. In response to the outcry regarding the potential impacts, the Cayman Islands Government (project supporting coalition) announced that they would be revising the cruise berthing facility design model. They stated that by placing two finger piers in deeper waters it would reduce the amount of dredging necessary as well as reduce the coral impact.

Another instance of policy-oriented learning by the project's supporting coalition included the revision of coral mitigation efforts. Once the preferred bidder was announced, some promotional material stated that they would aim for a 10:1 coral mitigation program, far greater than the original 1:1 coral mitigation claim. The goal was to replace ten times the amount of coral that would be lost, which the supporting coalition stated would mean that this project, while destroying naturally occurring reefs, would actually increase total coral as a result. This was a claim strongly contested by scientists and the opposition coalition based on a lack of scientific data showing this level of mitigation could even be possible given current technological constraints.

Another high-profile instance of policy-oriented learning occurred in November 2019 when a member of the Legislative Assembly moved seats and joined the project opposition. The legislative member did not approve of the government spending public tax money on the public information campaign for the project, arguing that there were too many conflicts of interests at play. Under several Freedom-of-Information requests, it was revealed that the government had spent over $230,000 of public money on the Support Our Tourism campaign. An example of a conflict of interest was found in the Minister of Tourism's family owning several waterfront shops that would stand to benefit from the placement of the port project, with passengers disembarking directly into their shopping center. The former premier, as part of the project supporting coalition, argued that this campaign was not a conflict of interest, but rather was well within the government's rights and that the legislator switching sides did not come as a surprise as that member had already expressed their reservations for the project.

Another instance of learning by the project's supporting coalition occurred during the lawsuit filed by a member of CPR Cayman. The legal challenges questioned the right to spend public money on the government's campaign, the set date for the referendum occurring so close to Christmas on December

19, 2019, and the neutrality of the wording of the question. In response to these challenges, the former Cayman Islands Government tweaked the referendum question. The original question for the referendum was: "Should the Cayman Islands continue to move forward with building the cruise berthing and enhanced cargo port facility?" After the legal challenges, the government amended the question to: "Should the Cayman Islands continue to proceed with building the cruise berthing and enlarged and refurbished cargo port facility?" (Ragoonath, 2021). However, this minor change did not dissuade the legal challenges made by the opposing coalition, which still proceeded and was successful in delaying the referendum into 2020.

Lastly, with the onset of the COVID-19 global pandemic in March 2020, the cruise berthing facility and referendum took a backseat. When the project was paused and a new general election was held, the former premier stated that the future of tourism should be less focused on cruise tourism and that once the borders reopen, the Cayman Islands might limit visitors to accommodate and not overwhelm the current systems they have in place. This was in contradiction of the former premier's prior beliefs as it was he who strongly advocated for the cruise berthing facility project, indicating strategic policy-oriented learning.

While the statements from the project opposition coalition stayed consistent, coalition members stated that they learned and gathered information from international scientists and each other that further vindicated their stance. Some stated that they were not a science-based group, and while the destruction of corals was alarming, it was difficult to argue against the mitigation efforts proposed. However, from prominent scientists in the United States and Australia, coalition members learned that what the Cayman Islands government was proposing was highly uncertain, with nothing of its scale having ever been successfully completed, information that was completely missing in the Government's Save Our Tourism public meetings, advertisements and campaign material. One respondent stated that understanding the science behind the mitigation efforts proposed was integral in the campaign and contesting the government, and that as learning increased, the intensity of the opposition increased as well since it became clear that once reefs were destroyed, it would not be simple, easy or affordable to "replant" them as Government and other project supporters insisted.

Furthermore, with the political victory inherent to triggering the people's initiated referendum, anti-port project coalition members "changed the political landscape" by reminding prominent political actors that consent was needed for a project of such magnitude to continue, and that suppressing opposing views, transparency or scientific experts would not lead to consent.

Finally, the opposition to the port project facilitated many more meetings and events than the coalition that supported the port project. The opposition coalition held meetings and events throughout the CPR Cayman campaign which began in 2018 and lasted through the pandemic in 2020. In October 2018, members of the opposition coalition enacted a series of public meetings and invited the Minister of Tourism multiple times; however, he was not able to attend these meetings. Between petition sign-up events, protests and rallies, the CPR Cayman group also hosted community meetings in January, May and October 2019. They also held an additional series of public meetings in November 2019, after the verification of the referendum was confirmed and shortly before the legal challenges against the Cayman Islands government. These events and meetings allowed for a high degree of policy-oriented learning among coalition members and the public. In contrast, the Cayman Islands Government (pro-port coalition), held two public meetings between 2015 and 2018, one of which was the review of the environmental impact assessment. When they announced the preferred bidder in July 2019, they held a press conference and discussed the plans for the cruise berthing facility project. Similarly to the anti-port project coalition, they also did a series of eight public meetings in November 2019. However, much to the disappointment of the public, there was limited dialogue between the Cayman Islands Government and public, as all questions had to be written and submitted during the meeting, and were selected by a government representative. This all resulted in a lack of learning opportunities between coalitions and sowed further mistrust and confusion, which eventually led to major policy change.

Implications for the Future

The significance of this case is the organization of a popular movement, CPR Cayman, to enact major policy change. Their success in collecting enough signatures to trigger the first people's-initiated referendum required the former Cayman Island Government to delay the project. Without the efforts of this coalition, it is likely that despite the public outcry, the government would have continued with the project. CPR Cayman also made an expensive effort to engage the public and elevated their concerns over the project to an international level, getting the attention of prominent news reporters, diving organizations and coral scientists to support their campaign.

The lack of good governance also contributed to CPR Cayman's success. CPR Cayman stated multiple times that the necessity to create their campaign was caused by the lack of information and transparency by the government. They believed there was not enough information on how the project would be financed, on the ecological impacts and the economic impacts from the destruction of coral reefs. The former government's actions, including removing the director of the Department of the Environment, enabled the opposing coalition to emphasize these shortcomings. Cayman is not the only place where events like these have transpired. For example, Lawer (2019) also found that dismissing the requests from local stakeholders led to litigations and delays of a port project in Ghana.

One of the most important takeaways from this case is the subject of science. The scientific technology to mitigate losses of coral reefs on a 10:1 in a way that ensures that all restored coral survives does not yet exist. Yet, this was promised to the public as the mitigation plan by the government, outsourced to a non-profit. The leaders within the government even claimed that this project would result in coral cover gained as opposed to lost by the Cayman Islands. The moral gray area where members of a regime assure the public about a technological fix to an environmental disturbance would not have been widely recognized were it not for the efforts of the opposition coalition pointing out the reality of the planned mitigation.

The Cayman Islands have only recently announced in March 2022 that they would be reopening their port to cruise ships again. However, CPR Cayman is still requesting for a referendum to ensure that the project won't be resumed as cruise tourism picks up again. There is disagreement among political leaders as to whether the approved referendum applies to the specific project described in this case, or to similar future projects. The current government administration has also not acknowledged CPR Cayman's request to hold the referendum.

This chapter presents a critical case study in which we observe pressure from non-governmental organizations drawing attention to scientific uncertainty and a lack of transparency over environmental impacts, resulting in major policy change. It gives us insight into how coalitions, through coordination and learning, can ensure that democratic processes such as people's-initiated referenda can occur. It also highlighted the need for much greater public participation, communication and transparency between the government and public. While local stakeholder actions, such as the mobilization of environmental non-governmental organizations, might be necessary to lead to policy change, it is important to remember that our governments, as trustees of our natural resources, and the environmental regulations in place should be sufficient to protect our precious resources, such as coral reefs.

References

Aichi Biodiversity Targets. (2020, February 5). Convention on biological diversity; secretariat of the convention on biological diversity. https://www.cbd.int/sp/targets/

Baird. (2015, February 5). *Proposed cruise berthing facility, grand Cayman environmental.* http://www.dlp.gov.ky/portal/pls/portal/docs/1/11910177.PDF

Boström-Einarsson, L., Babcock, R. C., Bayraktarov, E., Ceccarelli, D., Cook, N., Ferse, S. C. A., Hancock, B., Harrison, P., Hein, M., Shaver, E., Smith, A., Suggett, D., Stewart-Sinclair, P. J., Vardi, T., & McLeod, I. M. (2020). Coral restoration – A systematic review of current methods, successes, failures and future directions. *PLoS ONE, 15*(1). https://doi.org/10.1371/journal.pone.0226631

Boxall, J. (2021, February 5). The Cayman Islands financial services industry | Cayman 116 Resident. https://caymanresident.com/business/financial-services

Cayman Islands Environment Charter. (2001, February 8). https://www.amplifycayman.com/_files/ugd/1e07ee_8717b1c228104f12900aa5e4a6bb6588.pdf

Cayman Islands Government. (n.d.). *Finance & economy.* Retrieved July 11, 2021, from https://www.gov.ky/about-us/our-islands/finance-economy

Cayman News. (2019, January 29). *Stay-over spend eclipses cash from cruisers - Cayman Islands headline news: Cayman news service.* Caymans News Service. https://caymannewsservice.com/2019/01/stay-over-spend-eclipses-cruisers/

Department of Environment. (2009, February 8). *Cayman Islands national biodiversity action plan 2009.* https://cnslibrary.com/wp-content/uploads/Cayman-Islands-National-Biodiversity-Action-Plan-2009.pdf

Dunning, K. H. (2021). Adaptive governance: The proposed port expansion in the Cayman Islands and its impacts to coral reefs. *Marine Policy, 124*, 104248. https://doi.org/10.1016/j.marpol.2020.104248

Jenkins-Smith, H. C., Nohrstedt, D., Weible, C. M., & Ingold, K. (2018). The advocacy coalition framework: An overview of the research program. In *Theories of the policy process* (pp. 135–171). Routledge. https://doi.org/10.4324/9780429494284-5

Klein, M. (2019, November 25). Financing model analysis: Who is paying for the cruise berthing facility? *Cayman Compass.* https://www.caymancompass.com/2019/11/25/financing-model-analysis-who-is-paying-for-the-cruise-berthing-facility/

Lawer, E. T. (2019). Examining stakeholder participation and conflicts associated with large scale infrastructure projects: The case of Tema port expansion project, Ghana. *Maritime Policy and Management, 46*(6), 735–756. https://doi.org/10.1080/03088839.2019.1627013

López, A. M. (2021, July 5). *Inbound tourism volume in the Cayman Islands 2019.* Statista. https://www.statista.com/statistics/816372/cayman-islands-number-of-tourist-arrivals/

Matthews, G. (2020, August 14). *The road to tourism recovery.* Cayman Islands Government. https://www.gov.ky/news/press-release-details/the-road-to-tourism-recovery

McCarthy, D. (2018). *Carnival, royal Caribbean tell grand Cayman traffic will decrease without new pier.* TravelMarket Report. https://www.travelmarketreport.com/tmrarticle?ttl=Carnival-Royal-Caribbean-Tell-Grand-Cayman-Traffic-Will-Decrease-Without-New-Pier&fbclid=IwAR3B0N1tK_6l7aAW1xNI8MwRMmRMcD7F6h8nLl7vpTo9VaR2HqOB-ycg2B0#.W7a59Go0c7k.facebook

Ministry of Financial Services. (2021). *About Cayman.* https://www.mfs.ky/about-us/about-cayman/

Moses Kirkconnell: Why we need cruise berthing piers. (2018, September 20). *Cayman Compass.* https://www.caymancompass.com/2018/09/20/moses-kirkconnell-why-we-need-cruise-berthing-piers/

Ragoonath, R. (2021, May 6). Premier: No plans for a cruise berthing project - Cayman Compass. *Cayman Compass.* https://www.caymancompass.com/2021/05/06/premier

-no-plans-for-a-cruise-berthing-project/?fbclid=IwAR1z1BBCp0Hvoqiwue27ktWdS8kZeIcZQdsRqMVkixD4wW43_Fz5IGLltMU

The Cayman Islands Constitution Order, (2009, May 5). https://www.legislation.gov.uk/uksi/2009/1379/pdfs/uksi_20091379_en.pdf

The Marine Conservation Law, 1978 (2007 Revision, April 28), (2007). http://gazettes.gov.ky/portal/pls/portal/docs/1/11525201.PDF

The National Conservation Law, (2013, April 28). https://doe.ky/wp-content/uploads/2015/01/NationalConservationLaw-Es052014_web.pdf

Ware, M., Garfield, E. N., Nedimyer, K., Levy, J., Kaufman, L., Precht, W., Scott Winters, R., & Miller, S. L. (2020). Survivorship and growth in staghorn coral (Acropora cervicornis) outplanting projects in the Florida Keys National Marine Sanctuary. *PLoS ONE*, *15*(5), e0231817. https://doi.org/10.1371/journal.pone.0231817

Whittaker, J. (2021, April 28). Royal Caribbean boss "undecided" on Cayman cruise pier project. *Cayman Compass*. https://www.caymancompass.com/2021/04/28/royal-caribbean-boss-undecided-on-cayman-cruise-pier-project/?fbclid=IwAR3-lYn35YkPSy5DQ5wCdUXxz_uz-PHZMFdl- 9hopr2bV1EsfnwHWFb6cFE

Chapter 8

CULTURAL SERVICES OF REEFS: THE CASE OF THE CAYMAN ISLANDS MPAS AND WHAT WOULD BE LOST WITH A MAJOR INFRASTRUCTURE PROJECT

By Daniel Morris and Kelly Dunning

Acknowledgments: Data analysis and quality assurance of data provided by Research Assistants Catherine Cummings, Gregory Johnson, Danielle Oates, Vincent Rivers, Trinity Russo, Caroline Ward, Jordan Windham

What Are Ecosystem Services?

Just after my fifth birthday, my father took me fishing for the first time. He set my tiny Mickey Mouse fishing rod on the dock and did his best to explain to me the fundamentals of freshwater fishing. Hooks, bobbers and bait were the subjects at hand. It seemed very complicated, but I eagerly nodded my head as though I understood the lessons he was teaching me. I was certainly more interested in the excitement of pulling a live fish from the depths than learning the techniques needed to accomplish this feat.

He showed me how to carefully put a worm on the hook, and I was ready to go. I dropped the line in the water and expected a quick bite. Nothing. The minutes crept by, and it seemed the fish had much less interest than I did. My attention span waned as I sat waiting for something to bite. I began to look around. I heard the water lapping against the dock. The birds calling to one another from the shrubs and trees lining the pond. I smelled the warm Alabama breeze and admired the clouds lazily drifting overhead. There was so much happening, and yet nothing was happening all at once. As the morning turned to afternoon, it was obvious that the fish were not interested in our bait, and we left the pond without catching a thing.

While it's true that we weren't able to get a fish, all was not lost. Something else had stirred in my young mind. For the first time in my life, I felt connected to something more. I was only a tiny piece of something greater, something far more connected than I could comprehend at the time. Even though we didn't catch anything, I wanted more. I found value in the connectedness and the beauty of the Alabama ecosystem. I felt like a fisherman. I will always fondly remember this morning on the dock with my father as the first time I truly felt like I had an identity beyond that of a child.

While it is certainly true the ecosystem plays a vital part in providing resources to the humans that inhabit them, in this instance, it was the immaterial benefit that I found to be far more valuable. In other words, it was the experience itself over the fresh fish that could have been caught for a meal that was valuable. In this chapter, we will dive further into these types of benefits and show the importance of these cultural ecosystem services.

Ecosystem services are defined as the advantages or aid provided to humans from the ecosystem. Several scholars have refined this definition, and perhaps the most noteworthy explanation of ecosystem services can be found in the Millennium Ecosystem Assessment published in 2005. The authors define ecosystem services as "the benefits people obtain from ecosystems" (Millennium Ecosystem Assessment (Program), 2005, p. V). Within this broad definition, the benefits provided can be further grouped into four different types of benefits; provisioning services, regulating services, cultural services and supporting services. These four categories provide us with a reference point with which we can group almost any benefit provided to humans from the ecosystem. It also gives us a starting point to analyze the value these benefits provide to the human population.

The first category provided by the Millennium Ecosystem Assessment are provisioning services. These are defined as anything acquired or extracted from the ecosystem which benefits people such as food, water, timber, oils, natural gas and a multitude of other resources (Millennium Ecosystem Assessment (Program), 2005, p. V). Provisioning services provide living things, and especially humans, with the resources needed to sustain life. Each of us has undoubtedly experienced some benefit from the extraction of at least a few of these resources. How would we survive without food to sustain our bodies, or building materials to construct our homes? Since the first human stepped foot on the Cayman Islands, provisioning services such as food and water have played a fundamental role in sustaining human well-being. Life on the Caymans has been tied to the resources which humans can extract from the island, and more importantly, the waters surrounding the island.

The next category is regulating services. Regulating services are the benefits provided by nature that control or regulate natural processes. Pollination, erosion prevention and carbon storage are examples of regulating services (Millennium Ecosystem Assessment (Program), 2005, p. V). Many of the provisioning services, such as food and water would not be possible without regulating services such as pollination or water purification.

The third type of ecosystem services are supporting services. These services include any processes that allow the earth to provide for living things and the ecosystems in which they exist. Photosynthesis, soil formation and nutrient cycling are all examples of supporting services (Millennium Ecosystem Assessment (Program), 2005, p. V). The main distinction in differentiating supporting services from the other categories of ecosystem services are the key words "activities" or "processes." These actions build the foundation for all other ecosystem services.

The fourth and final type of benefit outlined in the Millennium Ecosystem Assessment, and the focus of this chapter, are cultural services. Cultural services are defined as the non-material benefits that aid in the betterment of human culture. This includes benefits from recreational activities, spiritual practices, acquisition of knowledge and even inspiration provided from the natural setting (Millennium Ecosystem Assessment (Program), 2005, p. V). Cultural services even provide individuals and groups with a sense of identity, from which further benefits such as unity and enthusiasm for the future can be derived.

Other scientific organizations have attempted to define ecosystem services as well. Of note, the Intergovernmental Science-Policy Platform on Biodiversity and Ecosystem Services (IPBES) utilizes the same definition ("the benefits people obtain from ecosystems") in their 2019 global assessment report on biodiversity and ecosystem services. IPBES categorizes these services into the same categories (provisioning, regulating, cultural and supporting services) as the Millennium Ecosystem Assessment (IPBES, 2019, p. 1038).

We chose this definition because it encompasses the wide range of benefits provided by the ecosystem and has been widely accepted among scientists and scholars. Almost any benefit, either material or immaterial, can be grouped into one of these categories. By categorizing these benefits, we can focus our efforts on discussing the merits and values associated with each category. It gives us a chance to disentangle the different ecosystem services from one another and present an informative analysis of each category. In keeping with the Millennium Ecosystem Assessment definition of ecosystem services and the categories used in this definition, in this chapter, we will simply define cultural ecosystem services as "the nonmaterial benefits people obtain from ecosystems through spiritual enrichment, cognitive development, reflection, recreation, and aesthetic experiences (Millennium Ecosystem Assessment (Program), 2005, p. 40)."

The Millennium Ecosystem Assessment and IPBES are not alone in attempting to group cultural ecosystem services together. Other scholars have grouped cultural ecosystem services into different categories with slightly different definitions; however, the services these functions provide are typically the same. Some different names for the same services include "cultural services" (Costanza et al., 1997, p. 254), "life-fulfilling functions" (Daily, 1999, p. 6), "information functions" (de Groot et al., 2002, p. 395), "amenities and fulfillment" (Boyd & Banzhaf, 2007, p. 623), "cultural and amenity services" (de Groot, 2010, p. 21), or "socio-cultural fulfillment" (Wallace, 2007, p. 241).

A majority of cultural ecosystem services definitions agree that the effect is intangible and often hard to quantify. Because they are hard to define and quantify, this category of ecosystem services has received less attention than the other categories when analyzing their importance and value as they relate to humans. Some scholars have argued that because these benefits are difficult to qualify, they should receive more consideration in future research (Milcu et al., 2013, p. 1). Because of this, cultural ecosystem services are often not valued appropriately or integrated into management plans (R. de Groot et al., 2005, p. 457). More often than not, cultural ecosystem services are not reflected by surrogate economic indicators such as real estate prices (Carpenter et al., 2009, page; Martín-López et al., 2009, p. 9).

Our Approach

To analyze our data we have chosen to use Fish et al.'s 2016 framework for cultural ecosystem services. Frameworks allow us to study abstract ideas using real-world data, and define concepts in concrete terms to compare them across different respondents, in our case, the people who care about the coral reefs of the Cayman Islands. We use two components of Fish et al. 2016's wider framework, the complete details of which are broken down in the Appendix for Chapter 9. First we focus on their *theory of cultural practices*. These cultural practices are activities that relate people to each other and the natural world. For example, in the Cayman Islands snorkeling and diving are very popular activities which provide several benefits to the local population. Fish et al. further break down cultural practices into four categories of activities that people engage in, which builds this connection. These categories include playing and exercising, creating and expressing, producing and caring, and gathering and consuming (Fish et al., 2016, p. 213).

Cultural practices and ecosystem services

Playing and exercising include activities which involve some type of physical interaction between people and the natural environment. These can be fitness or leisure activities that could be either individual pursuits or communal endeavors. Important examples of playing and exercising in the Caymanian context include free diving, snorkeling, SCUBA diving and relaxing on the beach. Out of 175 responses, 100 percent of respondents reported engaging in this type of activity while visiting the Caymanian reefs.

Creating and expressing activities are defined as those which involve the creation of representations of life, nature or any other artistic impression. This includes activities such as painting, sketching, writing, poetry, photography, music or storytelling. Customs, rituals or performances that emulate or are inspired by the natural environment are also included in this category of cultural practices. Photography is the most popular activity in this category reported in the Caymanian case.

The third type of cultural practice in the framework is *producing and caring*. These activities are defined as those which involve work or leisure activities related to conserving or managing the natural environment. This includes any employment activities that involve conservation or management of natural resources such as agriculture, fishing, environmental volunteering or scientific research. In the context of the Cayman Islands, fishing outside of the MPA was the most popular type of activity in this category.

The fourth and final type of cultural practice is *gathering and consuming*. These are defined as activities that may occur as either work or leisure endeavors that encompass collecting or consuming products from the natural world. This includes the consumption of local food, gathering of natural products such as wood products, and the consumption of local art, media or even performances. In the Cayman Islands, gathering of conch from the reef was the most popular activity in this category.

Cultural ecosystem benefits

Next we focus on Fish and colleagues' theory of cultural ecosystem benefits. This is best described as the dimensions of human well-being associated with cultural practices. These benefits are further broken down into three parts; identities, experiences and capabilities (Fish et al., 2016, p. 213) described in further detail later in this chapter.

Identities are the aspects of the natural world which enable people to connect with their environment to better understand their own existence and meaning. Identities include feelings of belonging, a sense of place, rootedness and spirituality. *Experiences* are events and the physical or emotional feelings people gain when interacting with the natural environment. Benefits associated with experiences include feelings of tranquility, inspiration, escape and discovery. *Capabilities* are the benefits from interacting with the natural environment which lead to an individual's ability to learn new skills to better their lives. Examples of capabilities include the gaining of skills or knowledge about the environment or one's self.

We selected a definition of cultural ecosystem services that focuses on centering the voices of stakeholders because cultural ecosystem services are created by the people ("co-produced"), so they're different wherever you go, and the thing that creates them is how people interact with ecosystems. For example, the intangible benefits of inspiration and connectedness that I experienced fishing as a youth wasn't something I simply plucked from the pond. They were formed within my mind, my interaction with the natural world, and synergized through my interactions with my father.

Understanding cultural ecosystem services can be difficult. The primary reason for understanding their values and importance is because of the intangibility of these services. Many of the benefits obtained from cultural ecosystem services are nondescript and instinctual (Kenter et al., 2011, p. 506). Oftentimes, observation of these benefits aren't as obvious or pronounced as provisioning and regulating services. They cannot be observed directly and we must utilize other indicators to gain an understanding of their value (Anthony et al., 2009, p. 4). Critics of the Millennium Ecosystem Assessment have argued that this definition of cultural ecosystem services does not clearly delineate the differences between the services ecosystems provide, the benefits associated with these services, and how these benefits should be valued (Chan et al., 2012, p. 1).

Potential Loss of Reefs to Infrastructure

As described in depth in Chapter 7, in the Cayman Islands, tourism is a major industry. Most of the tourists that visit the island arrive on cruise ships in George Town Harbour. In 2013 the Caymanian government announced that it would seek to expand its current cruise berthing facility in a project that would include construction of an additional pier, land reclamation and dredging operations. The government was concerned that the major cruise operators such as Royal Caribbean and Carnival Cruise would bypass the Cayman Islands in favor of ports with more accessible amenities.[1] After years of intense debate among citizens, the government and interest groups, the situation reached a tipping point. Opponents of the project gathered the signatures of over 25 percent of eligible Caymanian voters in order to prompt a people's initiated referendum in which the question of the expansion of the

1 The current cruise berthing facility is a tender port in which passengers are ferried between the ship and the island. The proposed expansion project would allow large cruise ships to dock directly at the piers, simplifying the movement of passengers on and off of the cruise ships.

cruise berthing facility would be put to a vote. After a delay due to the lack of a legal precedent of a referendum and procedural questions about how votes would be tallied, the vote was scheduled for 2020. However, due to the COVID-19 pandemic, the vote was indefinitely paused and the expansion project was shelved. The current Caymanian government has claimed that they will not pursue the project further; however, it is likely the same interest groups who initially pushed for the expanded infrastructure project will continue to advocate for a similar project in the future (Bailey et al., 2022, p. 7). Our research asks how cultural services would be impacted by a major infrastructure project that would result in a significant loss of coral reef ecosystems.

Cultural Services of Caymanian Reefs

Table 8.1 displays the most popular reef locations according to respondents living on the Cayman Islands, with the proposed port expansion project potentially impacting all of the most popular reefs through the negative impacts of dredging. All reefs in these systems fall under one or more statutory protections for coral reefs, including falling within an MPA and/or containing threatened and endangered species.

Of the four categories of cultural ecosystem benefits, experiences were the most important to Caymanians (80 percent of statements mentioned this category), followed by identities (25 percent of statements), capabilities (21 percent of statements), and other benefits that don't clearly fall into one of the other categories (11 percent of statements). Table 8.2 contains this breakdown. This suggests of primary importance to Caymanian coral reef stakeholders are services involving experiences to the greatest degree. An example was feeling a sense of wonder while diving on Eden Rock, a grotto-like cave system filled with tiny silvery fish in the summer.

Of the four concepts for cultural practices, playing and exercising were the most important (100 percent of statements mentioned this category), followed by producing and caring (21 percent of statements), gathering and consuming (18 percent of statements), and creating and expressing (2 percent of statements). Table 8.3 contains this breakdown. This suggests of primary

Table 8.1 Most popular reefs.

Location	*Percentage of Responses That Mention Site (*n*=176)*
Eden Rock	0.40
George Town Harbour	0.17
Macabuca	0.12

Table 8.2 Portions of statements containing each type of cultural ecosystem benefit.

Concept	*Portion of Statements with Concept Mentioned*
Identities	0.25
Experiences	0.80
Capabilities	0.21
Other	0.11

Table 8.3 Portions of statements containing each type of cultural ecosystem service.

Concept	*Portion of Statements with Concept Mentioned*
Playing and exercising	1.0
Creating and expressing	0.02
Producing and caring	0.21
Gathering and consuming	0.18

importance to Caymanian coral reef stakeholders are practices involving playing or exercising to the greatest degree, followed by producing and caring as well as gathering and consuming, which are nearly tied in terms of their relative importance. The following section provides examples of each type of statement from respondents to clarify the meaning.

Experiences on the reef

Cultural benefits related to experiences were the most important types of cultural services to respondents in this study. Statements touching on experiences were often focused on specific reefs that respondents liked to spend time on, such as Eden Rock, South Sound or the popular wreck dive of the USS *Kittiwake*, which according to the words of one respondent constitute "places of healing for the people." Experiencing the reef took place via glass-bottom boats, snorkeling, diving, accessing reefs via sailboat, and others suggesting these livelihoods are linked to the cultural benefits.

Many of the statements on experiences noted that Caymanian reefs are easy to get to, with access often coming directly from the shore. This means that not just those with access to a boat or funds to purchase a seat on a charter boat have access to Caymanian reefs. The following statement is representative of this concept:

> *Eden Rock is one of the most beautiful reefs. I remember diving through the tunnels and seeing the huge tarpon swim by. I felt one with nature, and awe inspired by the*

incredible beauty. I also love that it is swimmable from the shore. It's one of the best snorkel spots on this side of the island.

Reefs as Caymanian identity

Cultural services related to Caymanian identity followed three patterns with statements having to do with: (1) who Caymanians are as a culture, (2) how reefs are important parts of cherished memories going all the way back to childhood, and (3) how reefs are linked to spirituality. First, respondents saw coral reefs as an essential part of what it means to be a Caymanian person. For example, respondents would make statements about how they knew that reefs were symbolic of their "home" of the Cayman Islands, or that Caymanians have something special or rare in their reefs and this is an essential part of Caymanian culture. This sentiment is evidenced in the following statement:

I enjoyed knowing this is my home and we have something special that many rarely (if ever) have the opportunity to experience, and yet this underwater world that is outside my doorstep [and] is something I can experience at any time- with respect and appreciation.

The second type of statement about reefs being a part of Caymanian identity include ones that reminisce about spending time on reef from early childhood through adulthood, and how these experiences shaped their current lives as adults.

I have been snorkeling all these reefs since I was 2 and diving since I was 10. They are still and always will be fascinating and dear to me. Each time I snorkel or dive it is like it is the first time yet it feels like I am visiting my old neighborhood again! I never get tired or bored even though I know each part of these reefs intimately. My best friend and I made a pact when I was 17 that we will still be diving these reefs when we are in wheelchairs!

The third and final component of identity-based cultural benefits has to do with finding a form of spirituality on the reefs, broadly defined. These statements often touched on ideas of divinity or God. Take for example the following representative statement: "In awe of the creative power of God. The colors and varieties of sea life gives me a sense of peace and wonder." Another example of spirituality among respondents includes the following statement:

I view my world through a secular lens but if there were any spiritual experiences I could credit to developing both my soul and psyche it would be due to the being of a human—simply being—amongst the rich tapestry of biodiversity and Creation that are the reefs of the Cayman Islands.

Learning on the reef

Statements involving capabilities were made by respondents acquiring knowledge or engaged in conservation for their jobs or as volunteers. Often, statements focused on teaching one's children about biodiversity and conservation by bringing them to the reefs of the Cayman Islands. The words of the following respondent are characteristic of this type of cultural benefit:

> *I had my two children holding each others' hands as we snorkeled. We would stop and dive down to look under huge coral heads and see the lobsters' antennae and fish hovering there. Once a large moray eel came out and surprised us. It was like introducing them to a magic world—pointing at all the extraordinary creatures and the joy when they spotted something amazing.*

The importance of transmitting knowledge to future generations was valued by respondents on its own, but also as a mechanism for conservation. Educating young children about the importance of biodiversity conservation is one way to ensure that conservation is a priority for younger generations.

Playing and exercising on the reef

The cultural practice of playing and exercising was the most important cultural practice to the Caymanian people. 100% of respondents reported engaging in some sort of playing or exercising activity while visiting the reefs around the islands. Examples of these practices from respondents include snorkeling, diving and relaxing on the beach. This sentiment is shown in the words of this Caymanian respondent:

> *I have dived the Georgetown reefs so many times! The last time I dived Eden Rock when the silversides were in was the most magical ever! I spent an entire two hours underwater, most of that time was spent just lying on the sandy bottom in one of the tunnels watching the silversides. I was absolutely mesmerized!*

The importance of recreational fishing

The cultural practice of producing and caring was the second most reported practice by Caymanian respondents (21 percent of responses). The most popular practice reported pertained to fishing in or around the islands for species like grouper and snapper. The following response shows just how pronounced this practice is for the Caymanian people:

As we are a premium dive and tourist destination our pristine waters and the beautiful reefs and supporting marine life are of paramount importance. Many locals livelihood and even food source depend on a healthy marine environment so this needs to be protected by all means necessary.

Local taste for shellfish

Gathering and consuming was the third most reported cultural practice reported by Caymanian respondents (18 percent of responses). Of note, respondents generally reported gathering conch and lobster from the reef as an example of a gathering and consuming cultural practice. To illustrate the importance of this practice, one Caymanian responded, "The reefs are life-sustaining to the island [...] Not just to the underwater ecosystem, but the reefs define Cayman. they are not an accessory."

Losses of Cultural Services with Major Infrastructure Projects

The previous section has shown that with negative ecological impacts to popular sites such as Eden Rock and the reefs of the Georgetown Harbour accompany a loss of ecosystem services. These services include experiences like accessing reefs via glass-bottom boat, Caymanian identity which links the reefs to Cayman way of life and potential teaching and learning about biodiversity. People's ability to exercise, play, fish and harvest shellfish will be lost if major infrastructure projects are allowed.

We asked respondents if they would still visit these sites for cultural reasons if the quality and clarity of the water were degraded. Opinions were split; however, 41 percent of respondents reported they would be less likely to visit the reefs (Figure 8.1). Additionally, we asked respondents about the role of coral reefs in Caymanian culture (figure 8.2) and identity (Figure 8.3). Almost all respondents were in agreement that the reefs were an important part of Caymanian culture, with over 91 percent responding that they "Strongly Agree" with the sentiment, and 85.7 percent "Strongly Agree" that the reefs were an important part of Caymanian Identity.

Additionally, Caymanians felt very strongly about the potential loss of the reefs around the islands. When asked, "If the coral reefs of the Cayman Islands were to be destroyed, how would you feel?" Respondents used powerful, emotional words like devastation, sadness and betrayal when describing their feelings. The following statements characterize a majority of the responses Caymanians provided to this question: "Devastated. It would be hard for me to look my son in the eye. It would be difficult to show him pictures of what

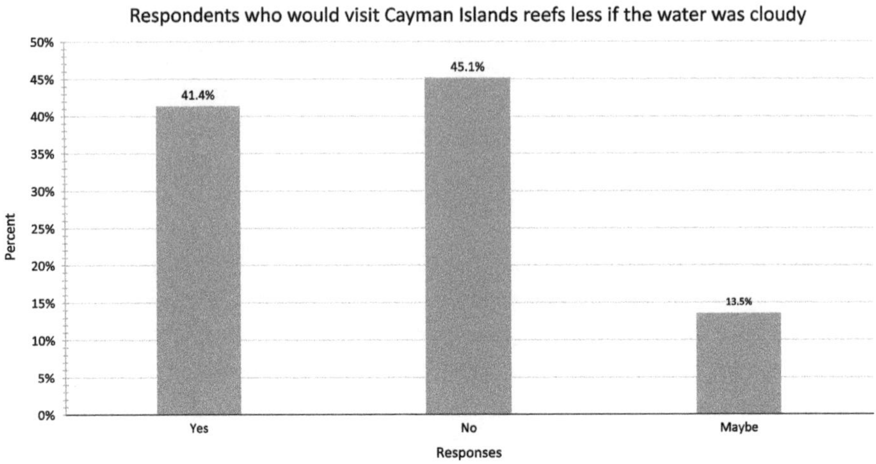

Figure 8.1 Responses to the question: "If the water was cloudy in the Georgetown Harbour, would you visit the reefs (Eden Rock, Devil's Grotto) less?"

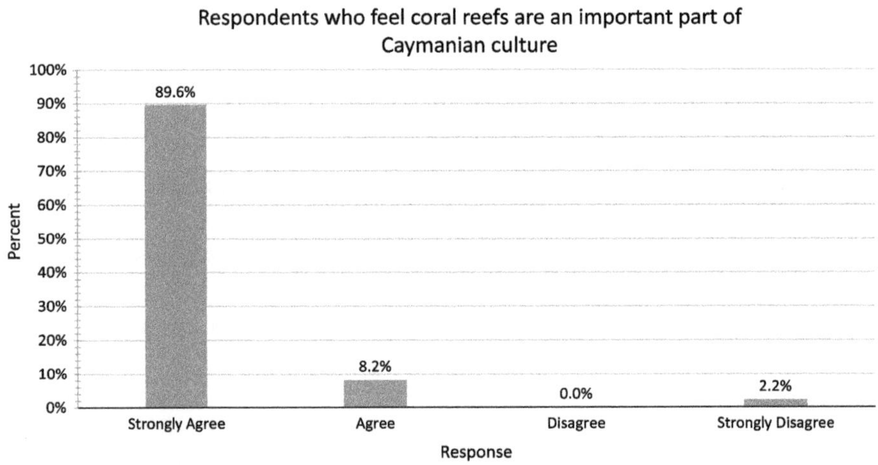

Figure 8.2 Respondents agree with the statement, "Coral reefs are an important part of Caymanian culture."

was once there." Another respondent said the following: "[I would feel] Very sad and distraught. The one thing that made us so unique has now gone and we have nothing left to boast about being the best island in the Caribbean." Additional responses include the following representative statements:

> *Heartbroken. It makes me want to cry thinking about it, actually. Destroying these places in the harbour would not only affect Caymanians, but those of use who have*

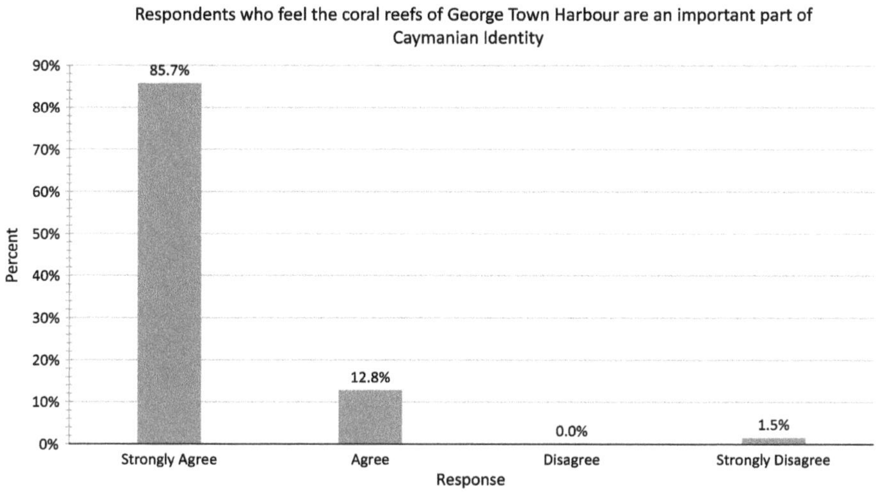

Figure 8.3 Respondents agree with the statement, "The coral reefs of George Town Harbour (Eden Rock, Devil's Grotto) are an important part of Caymanian identity."

> *grown to love this island like our second home. Cayman is UNIQUE (sic). We don't need another harbour town cruise stop. We need the unique beauty and splendor of the natural creation in Cayman. To undo this would be a travesty.*
>
> *I would feel very sad that this important part of Cayman's culture has been lost, and also very angry that our greedy politicians were willing to sacrifice George Town Harbour, and very likely Seven Mile Beach, instead of preserving it for our children, grandchildren and all future generations of Caymanians.*

Conclusion

While many cultural ecosystem services may be intangible or hard to quantify, this research should help bridge the gap in appropriately assessing the value these resources provide to the Caymanian people, beyond the provisioning, supporting and regulating services of coral reefs. Without assessing the cultural benefits, enabled by cultural practices in the Cayman Islands, we do not get a clear picture of the full value of these resources and the importance they hold for the people who rely on these resources every day. Consideration of the cultural values of an ecosystem should be carefully analyzed by decision-makers when considering trade-offs related to the alteration or loss of a specific resource, such as the potential loss of coral reef cover due to the proposed cruise berthing facility expansion in the Cayman Islands. This case study provides us with a clear example of how stakeholders opinions related to a resource hold value and matter to the user groups who rely on these resources. Much like my experience fishing as a young man, there is much to gain from properly managing these resources and protecting them for the benefit of future generations.

References

Anthony, A., Atwood, J., August, P., Byron, C., Cobb, S., Foster, C., Fry, C., Gold, A., Hagos, K., Heffner, L., Kellogg, D., Lellis-Dibble, K., Opaluch, J., Oviatt, C., Pfeiffer-Herbert, A., Rohr, N., Smith, L., Smythe, T., Swift, J., & Vinhateiro, N. (2009). Coastal lagoons and climate change: Ecological and social ramifications in U.S. Atlantic and gulf coast ecosystems. *Ecology and Society*, *14*(1). https://doi.org/10.5751/ES-02719-140108

Bailey, S., Dunning, K. H., Morris, D., & Williamson, R. D. (2022). How narratives shape policy: Lessons learned from port projects adjacent to coral reefs in Florida and the Cayman Islands. *Marine Policy*, *144*, 105233. https://doi.org/10.1016/j.marpol.2022.105233

Boyd, J. & Banzhaf, S., 2007. What are ecosystem services? The need for standardized environmental accounting units. *Ecological Economics*, *63*(2–3), pp.616–626.

Carpenter, S. R., Mooney, H. A., Agard, J., Capistrano, D., DeFries, R. S., Díaz, S., Dietz, T., Duraiappah, A. K., Oteng-Yeboah, A., Pereira, H. M., Perrings, C., Reid, W. V., Sarukhan, J., Scholes, R. J., & Whyte, A. (2009). Science for managing ecosystem services: Beyond the millennium ecosystem assessment. *Proceedings of the National Academy of Sciences*, *106*(5), 1305–1312. https://doi.org/10.1073/pnas.0808772106

Chan, K. M. A., Satterfield, T., & Goldstein, J. (2012). Rethinking ecosystem services to better address and navigate cultural values. *Ecological Economics*, *74*, 8–18. https://doi.org/10.1016/j.ecolecon.2011.11.011

Costanza, R., d'Arge, R., De Groot, R., Farber, S., Grasso, M., Hannon, B., Limburg, K., Naeem, S., O'neill, R.V., Paruelo, J., & Raskin, R. G., 1997. The value of the world's ecosystem services and natural capital. *Nature*, *387*(6630), pp.253–260.

de Groot, D., 2010. Protecting natural capital for human wellbeing and sustainable development. Science for Environment Policy, DG Environment News Alert, Special Issue Ecosystem Services, 1.

de Groot, R., van de Berg, A., & Amelung, B. (2005). Cultural and amenity services. *Ecosystems and Human Well-Being; Volume 1 Current State and Trends: Findings of the Condition and Trends Working Group*, p. 455.

Fish, R., Church, A., & Winter, M. (2016). Conceptualising cultural ecosystem services: A novel framework for research and critical engagement. *Ecosystem Services*, *21*, 208–217. https://doi.org/10.1016/j.ecoser.2016.09.002

IPBES. (2019). *Global assessment report on biodiversity and ecosystem services of the intergovernmental science-policy platform on biodiversity and ecosystem services.* Zenodo. https://doi.org/10.5281/zenodo.6417333

Kenter, J. O., Hyde, T., Christie, M., & Fazey, I. (2011). The importance of deliberation in valuing ecosystem services in developing countries—Evidence from the Solomon Islands. *Global Environmental Change*, *21*(2), 505–521. https://doi.org/10.1016/j.gloenvcha.2011.01.001

Martín-López, B., Gómez-Bagghetun, E., González, J., Lomas, P., & Montes, C. (2009). The assessment of ecosystem services provided by biodiversity: Re-thinking concepts and research needs. In *Handbook of nature conservation* (pp. 261–282). Nova Science Publishers.

Milcu, A., Hanspach, J., Abson, D., & Fischer, J. (2013). Cultural ecosystem services: A literature review and prospects for future research. *Ecology and Society*, *18*(3). https://doi.org/10.5751/ES-05790-180344

Millennium Ecosystem Assessment (Program). (Ed.). (2005). *Ecosystems and human well-being: Synthesis.* Island Press.

Wallace, K. J., 2007. Classification of ecosystem services: Problems and solutions. *Biological conservation*, *139*(3–4), 235–246.

Chapter 9

HOW CORAL REEFS ARE PLACED ON THE DECISION-MAKING AGENDA

By Amanda Alva, Kelly Dunning, Ryan Williamson

Data analysis provided by Research Assistant Kampol Pannoi

In 2021, a group of prominent Democratic and Republican lawmakers from the bicameral U.S. legislature, or Congress, came together to call for new policy to defend American coral reefs from environmental change. In the words of Republican Senator Marco Rubio, "I saw the devastated condition of our coral reefs firsthand when touring the Florida Keys National Marine Sanctuary, and I promised a comprehensive response." His perceptions on stressors to reefs and the need for their proposed policy, the Restoring Resilient Reefs Act, was nearly indistinguishable from his Democratic colleagues. Given how polarized American politics have been for the past decade, it is important to understand why lawmakers come together and agree upon the urgent need to protect coral reefs from stressors? Is there something inherently special to coral reefs that causes elected leaders to put aside typical partisan grudges and work together to ensure laws are passed to guarantee American reefs survive the era of climate change?

This chapter will review the history of coral reef policy in the U.S. Congress, including a focus on policies that had congressional support from both political parties, and a specific case of reauthorizing the fundamental coral reef conservation law in the United States. We argue that understanding why legislators decide to place coral reefs on the policy-making agenda is important for biodiversity conservation, but may also be a source of common ground over more contentious issues, such as the passing of climate change legislation. Our results can help interest groups message the importance of biodiversity and climate policy to lawmakers in a way that appeals to a range of political ideologies.

History of Coral Reef Policy in the U.S. Congress

The federal government of the United States is broken up into three branches of government: the Executive Branch, which is where the president and federal agencies carry out laws; the Legislative Branch, where laws are made in Congress; and the Judicial Branch, where laws are judged based on whether they follow the U.S. constitution. Congress is made up of two chambers or houses, the House of Representatives and the Senate. The House of Representatives is the lower house, which is generally characterized by early career politicians with more frequent elections every two years who represent a district within a U.S. state based on population. The Senate is the upper house, generally characterized by more experienced politicians with six-year election cycles that represent their entire state. Legislators in both chambers introduce bills, or draft laws, that they want to see become official U.S. laws.

To manage the volume and complexity of congressional bills introduced in the House of Representatives and the Senate, there are committees put in place to discuss and amend bills, reach out to experts for testimony, and hold hearings on issues (U.S. Senate, 2021). These committees[1] will hold hearings and public meetings, edit the bills, and vote on whether to report the bill to the floor of the respective chamber for a debate and vote. If it passes through the House of Representatives and then the Senate, it goes onto the president's desk where s/he can veto[2] or sign the bill to become a law.

Coral reef legislation in congress

Coral reef related laws in the United States initially focused on the mitigation of stressors for coral reefs, such as sediment runoff from coasts, and now focuses on establishing MPAs and coral reef restoration (Richmond et al., 2007). The earliest laws, the Rivers and Harbors Act of 1899 and the Fish and Wildlife Coordination Act of 1958, provided legal authorization to U.S. agencies for modifying waterways for navigation, before coral reefs were valued ecologically (Richmond et al., 2007).

---

1 The Authorizing Committees who provide insight on bills that introduce new programs include the House of Representatives Science, Space & Technology Committee, Natural Resource Committee, and Agriculture Committee and the Senate Commerce, Science & Transportation Committee, Energy and Natural Resources Committee, and Agriculture Committee (U.S. Senate, 2020).

2 A veto is essentially an action where the president refuses to sign a bill to become law and it is discarded and must start the process over again if legislators want to see it become law.

The Coastal Zone Management Act created a federal grant program for coastal states to plan and implement programs to promote effective coastal management of ecosystems (Richmond et al., 2007). The Endangered Species Act is one of the most important and long-standing conservation policies which provides protections and limitations for iconic species that make up coral reef ecosystems, including several species of coral (Richmond et al., 2007). The Clean Water Act of 1977 grants authority to the Environmental Protection Agency to regulate pollution in waterways, which is one of the most long-standing stressors to coral reef ecosystems due, specifically increased nutrient inputs from surface runoff (Richmond et al., 2007). The Magnuson-Stevens Fishery Conservation and Management Act requires science-based plans for managing key fisheries, including those that are located or spend parts of their lives on coral reefs (Richmond et al., 2007). The National Environmental Policy Act of 1969 was put in place to establish environmental impact standards for major infrastructure projects like bridges or piers to lessen environmental damage to ecosystems (Richmond et al., 2007). The National Marine Sanctuaries Act, National Park Service Organic Act, and National Wildlife Refuge System Administration Act are the basis for creating protected areas, including MPAs, under federal jurisdiction in the United States; and there are many aimed at protecting coral reef ecosystems (Craig, 2000).

The most important policy for the conservation of coral reefs is the Coral Reef Conservation Act of 2000, which funds coral reef conservation projects, such as coral reef nurseries or scientific research, within U.S. resource management agencies. The Coral Reef Conservation Act also led to the creation of the Coral Reef Conservation Program (Craig, 2000). The Coral Reef Conservation Program was the first of its kind in the United States and provided the framework for the decades of coral reef research, conservation and restoration work funded and completed by the federal government to protect these valuable ecosystems (Craig, 2000).

To date, there has been no attempt at categorizing all of the coral reef conservation policies passed by Congress. This is important to do because understanding the different types of policies allows advocacy groups and scientists to demand further action and similar policies or improvements to what already exists. We argue that there are four types of coral reef conservation policies that Congress has passed: those enabling the federal government to actively intervene on the reef resource for conservation, those furthering an existing federal conservation program, those funding science and those aimed at reducing reef stressors. We describe the four types of policy in the following sections and then determine why lawmakers support policy and place them on the decision-making agenda. Understanding this case is essential to understanding how laws that protect precious ecosystems like reefs are made, and perhaps how they can model how climate policy can be made in the future. The following sections present our analysis on coral reef conservation policy in the United States with methods and data collection information contained in the Appendix for Chapter 8.

Bipartisan coral reef conservation bills (1988–2021)

Bipartisan policies are those that are supported by two or more political parties in the United States. This is often in the scope of the two major political parties: Republicans, who support more conservative policies, and Democrats, who support more liberal policies. Since the 1980s, environmental policy has become polarized in the U.S. Congress between these political parties, with Democrats championing environmental policy and Republicans refusing to support environmental policy arguing that excess regulations impede economic growth (Smith, 2004). Bipartisanship is important because without it, environmental policies will not be passed through Congress (Kustic, 2020; Waldeck & Buck, 2001).

We found that there are four categories of bipartisan coral reef bills: (1) bills that enable the federal government to do something on the coral reef resource itself (e.g., intervening in the ecosystem for conservation), (2) bills that furthered a federal conservation program for coral reefs, (3) bills that further funding, ideas or science for reef conservation (as the primary focus of the bill), (4) bills that addressed specific stressors for coral reef ecosystems.

Bills that enabled the federal government to do something on the coral reef resource itself for conservation included authorization for creating MPAs, mandating federal management and conservation of coral reefs and creating rules for areas with coral reefs (Table 9.1). All attempts to reauthorize the Coral Reef Conservation Act of 2000,[3] an executive order that mandated the federal protection and conservation of coral reef ecosystems, were included in this section.

The Bunitan Tasi Act was included because it called for the enhancement of natural resource management, including activities to improve conservation of coral reef ecosystems on Guam. The Northwestern Hawaiian Islands National Marine Refuge Act of 2005 was included because it established an MPA for the conservation of coral reef ecosystems in the Northwestern Hawaiian Islands. The Bottom Trawl and Deep Sea Coral Habitat Act

3 The Coral Reef Conservation and Restoration Partnership Act of 2000, Coral Reef Conservation Act of 1997, Coral Reef Stewardship Act and Coral Reef Environmental Act were all precursors to the Coral Reef Conservation Act of 2000.

Table 9.1 Bills that enabled the federal government to do something on the coral reef resource itself for conservation.

Bill Name	*Congress*	*Year*	*Purpose*	*Status*
H.R. 160 Restoring Resilient Reefs Act of 2021	117	2021–2022	To reauthorize the Coral Reef Conservation Act of 2000, among other purposes.	Introduced
H.R. 6738 Coral Reef Conservation Reauthorization Act of 2020	116	2019–2020	To reauthorize the Coral Reef Conservation Act of 2000, among other purposes.	Introduced
H.R. 4160 Restoring Resilient Reefs Act of 2019	116	2019–2020	To reauthorize the Coral Reef Conservation Act of 2000, among other purposes.	Introduced
H.R. 5996 Coral Reef Conservation and Reauthorization Act of 2018	115	2017–2018	To reauthorize the Coral Reef Conservation Act of 2000, among other purposes	Introduced
H.R. 5771 Conserving Our Reefs and Livelihoods Act of 2016	114	2015–2016	To reauthorize the Coral Reef Conservation Act of 2000, among other purposes.	Introduced
H.R. 860 Coral Reef Conservation Act Reauthorization and Enhancement Amendments of 2009	111	2009–2010	To reauthorize the Coral Reef Conservation Act of 2000, among other purposes.	Passed House
H.R. 4493 Bunitan Tasi Act	111	2009–2010	To provide for the enhancement of visitor services, fish and wildlife research, and marine and coastal resource management on Guam related to the Marianas Trench Marine National Monument, and for other purposes.	Introduced
H.R. 1205 Coral Reef Conservation Amendments Act of 2007	110	2007–2008	To reauthorize the Coral Reef Conservation Act of 2000, among other purposes.	Passed House
H.R. 4788 Coral Reef Conservation Amendments Act of 2006	109	2005–2006	To reauthorize the Coral Reef Conservation Act of 2000, among other purposes.	Introduced

(Continued)

Table 9.1 (Continued)

Bill Name	*Congress*	*Year*	*Purpose*	*Status*
H.R. 2376 Northwestern Hawaiian Islands National Marine Refuge Act of 2005	109	2005–2006	To designate the Northwestern Hawaiian Islands National Marine Refuge, and for other purposes.	Introduced
H.R. 3778 Bottom Trawl and Deep Sea Coral Habitat Act	109	2005–2006	To establish ocean bottom trawl areas in which trawling is permitted, to protect deep sea corals and sponges, and for other purposes.	Introduced
H.R. 4897 Deep Sea Coral Protection Act	108	2003–2004	To protect deep sea corals and sponges, and for other purposes.	Introduced
H.R. 2654 Rigs to Reefs Act of 2003	108	2003–2004	To amend the Outer Continental Shelf Lands Act to direct the secretary of the Interior to issue regulations under which the secretary may authorize use of a decommissioned offshore oil and gas platform for culture of marine organisms, an artificial reef or scientific research, and for other purposes.	Introduced
H.R. 3919 Coral Reef Conservation and Restoration Partnership Act of 2000	106	1999–2000	To provide assistance for the conservation of coral reefs, to coordinate Federal coral reef conservation activities and for other purposes.	Introduced
H.R. 2233 Coral Reef Conservation Act of 1997	105	1997–1998	To assist in the conservation of coral reefs.	Passed House
H.R. 4483 Coral Reef Stewardship Act	102	1991–1992	To protect and promote stewardship of coral reef ecosystems.	Introduced
H.R. 4537 Coral Reef Environmental Act	102	1991–1992	The purpose of this Act is to promote the sustainable use of coral reef ecosystems and the conservation of the biological resources contained within them.	Introduced
H.R. 3719 Florida Keys Protection Act	101	1989–1990	To establish the Florida Keys National Marine Sanctuary, and for other purposes.	Passed House

requires NOAA to explore, identify, research and map the locations of deep sea corals and deep sea sponges in order to designate a Coral Habitat Conservation Zone for the protection of seafloor habitat. The Deep Sea Coral Protection Act also required NOAA to research and map deep sea corals and sponges along with prohibiting the use of specific fishing gear in any Coral Management Area and Coral Study Area.

In the face of the increasing use of installing artificial reefs and outplanting coral as a form of coral restoration, the Rigs to Reefs Act of 2003 was introduced to issue regulations for the National Oceanic and Atmospheric Administration to authorize the use of a decommissioned offshore oil and gas platform for artificial reefs. The Florida Keys Protection Act was included because it designated the Florida Keys National Marine Sanctuary under the Marine Protection, Research, and Sanctuaries Act of 1972.

Bills that further a federal conservation program included programs like debt for nature swaps and various management programs (Table 9.2). This included all reauthorizations of the Tropical Forest Conservation Act of 1998 and Coral Reef and Coastal Marine Conservation Acts which would initiate funding for debt for nature swap programs, where developing countries would pledge to enact conservation activities for coral reefs in their own country to alleviate debt owed to the United States (Alagiri, 1992; McFarland, 2018; Patterson, 1990).

The Oceans Conservation, Education, and National Strategy for the 21st Century Act established a national policy to protect, maintain and restore the health of marine ecosystems and required NOAA to use ecosystem-based management, or management that considers an entire ecosystems needs versus one species, to fulfill its responsibilities with respect to the conservation of marine mammals, protected species, coral reefs and other living marine resources.

The Coastal and Great Lakes Communities Enhancement Act was included because it supports projects to address the immediate and long-term degradation or loss of coral and coral reefs in response to bacteria, fungi, viruses, increased sea surface temperatures, ultraviolet radiation and pollutants. The Coastal Zone Act Reauthorization Amendments of 1990 was introduced to improve the management of coasts and enhance environmental protection of coastal ecosystems, including coral reef resources, by reauthorizing and amending the Coastal Zone Management Act of 1972. Its goal was to revitalize the federal coastal management by enhancing Federal leadership and technical and financial assistance in support of improved coastal ecosystem management at regional, state and local levels.

Bills that further funding, ideas or science for reef conservation as a main component were also considered (Table 9.3). These included the Offshore

Table 9.2 Bills that further a federal conservation program.

Bill Name	*Congress*	*Year*	*Purpose*	*Status*
H.R. 241 Tropical Forest and Coral Reef Conservation Reauthorization Act 2021	117	2021–2022	To reauthorize the Tropical Forest and Coral Reef Conservation Act of 1998.	Passed House
H.R. 7954 Tropical Forest and Coral Reef Conservation Reauthorization Act of 2020	116	2019–2020	To reauthorize the Tropical Forest and Coral Reef Conservation Act of 1998.	Passed House
H.R. 729 Coastal and Great Lakes Communities Enhancement Act	116	2019–2020	To amend the Coastal Zone Management Act of 1972 to authorize grants to Indian Tribes to further achievement of Tribal coastal zone objectives, and for other purposes.	Passed House
H.R. 6982 Tropical Forest Conservation Reauthorization Act of 2018	115	2017–2018	To reauthorize the Tropical Forest and Coral Reef Conservation Act of 1998.	Introduced
H.R. 52 Tropical Forest and Coral Conservation Act	111	2009–2010	To reauthorize the Tropical Forest and Coral Reef Conservation Act of 1998.	Introduced
H.R. 2185 To Amend the Tropical Forest Conservation Act of 1998	110	2007–2008	To amend the Tropical Forest Conservation Act of 1998 to provide debt relief to developing countries that take action to protect tropical forests and coral reefs and associated coastal marine ecosystems, to reauthorize such Act through fiscal year 2010 and for other purposes.	Passed House
H.R. 1996 Coral Reef and Coastal Marine Conservation Act of 2005	109	2005–2006	To amend the Foreign Assistance Act of 1961 to provide for debt relief to developing countries that take action to protect critical coral reef habitats.	Introduced
H.R. 1721 Coral Reef and Coastal Marine Conservation Act of 2003	108	2003–2004	To amend the Foreign Assistance Act of 1961 to provide for debt relief to developing countries who take action to protect critical coral reef habitats.	Introduced
H.R. 2272 Coral Reef and Coastal Marine Conservation Act of 2001	107	1991–2002	To amend the Foreign Assistance Act of 1961 to provide for debt relief to developing countries who take action to protect critical coral reef habitats.	Passed House
H.R. 4450 Coastal Zone Act Reauthorization Amendments of 1990	101	1989–1990	To improve management of the coastal zone and enhance environmental protection of coastal zone resources, by re-authorizing and amending the Coastal Zone Management Act of 1972, and for other purposes.	Introduced

Table 9.3 Bills that further funding, ideas or science for reef conservation.

Bill Name	*Congress*	*Year*	*Purpose*	*Status*
H.R. 1689 Offshore Wind for Territories Act	117	2021–2022	To amend the Outer Continental Shelf Lands Act to apply to territories of the United States, to establish offshore wind lease sale requirements, to provide dedicated funding for coral reef conservation and for other purposes.	Introduced
H.R. 3384 Coral Reef Sustainability Through Innovation Act of 2019	117	2021–2022	To authorize Federal agencies to establish prize competitions for innovation or adaptation management development relating to coral reef ecosystems, and for other purposes.	Introduced
H.R. 1014 Offshore Wind for Territories Act	116	2019–2020	To amend the Outer Continental Shelf Lands Act to apply to territories of the United States, to establish offshore wind lease sale requirements, to provide dedicated funding for coral reef conservation and for other purposes.	Introduced
H.R. 6665 Offshore Wind for Territories Act	115	2017–2018	To amend the Outer Continental Shelf Lands Act to apply to territories of the United States, to establish offshore wind lease sale requirements, to provide dedicated funding for coral reef conservation and for other purposes.	Passed House
H.R. 4811 Coral Reef Sustainability Through Innovation Act of 2016	114	2015–2016	To authorize Federal agencies to establish prize competitions for innovation or adaptation management development relating to coral reef ecosystems, and for other purposes.	Introduced
H.R. 2685 Climate and Ocean Research and Coordination Act of 2009	111	2009–2010	To establish a National Oceanic and Atmospheric Administration and a National Climate Enterprise, and for other purposes.	Introduced
H.R. 701 Conservation and Reinvestment Act	106	1999–2000	To provide Outer Continental Shelf Impact Assistance to State and local governments, to amend the Land and Water Conservation Fund Act of 1965, the Urban Park and Recreation Recovery Act of 1978, and the Federal Aid in Wildlife Restoration Act to establish a fund to meet the outdoor conservation and recreation needs of the American people, and for other purposes.	Passed House
H.R. 5617 Oceans Act of 1992	102	1991–1992	Amends the Marine Protection, Research, and Sanctuaries Act of 1972 to modify: (1) various definitions; (2) the factors to be considered, and the procedures to be used, in designating sanctuaries; (3) provisions relating to certain access and use rights; and (4) reporting requirements.	Became Law

Wind for Territories Act, which would create a dedicated coral reef fund from offshore wind revenue, and the Conservation and Reinvestment Act, which would allocate funds for the conservation, restoration, enhancement or protection of coral reefs. The Coral Reef Sustainability Through Innovation Act of 2019 and Coral Reef Sustainability Through Innovation Act of 2016 outlined prize competitions for management agencies for any program that innovated ways to understand, research or monitor coral reef ecosystems, or to develop management or adaptation options to preserve, sustain and restore coral reef ecosystems. The Climate and Ocean Research and Coordination Act of 2009 would have established a National Climate Enterprise to advance climate science in the United States to forecast natural and human-induced climate, climate change and climate variability to effectively carry out the intent of various conservation policies, including the Coral Reef Conservation Act of 2000.

Some stressors for coral reef organisms that policies address include hurricanes, pollution, fisheries and climate change (Table 9.4). Hurricanes physically damage coral reef structures through wave action (Wilkinson & Souter, 2008). Various versions of the Hurricane Research Initiative Act included research objectives to understand mitigation tools to decrease hurricane impacts on coastal environments, including coral reefs.

Cruise ships are a source of nutrient pollution to coral reef ecosystems. The Clean Cruise Ship Acts introduced in 2004, 2005, 2008 and 2009 would have established national standards for discharges from cruise vessels to take into account marine life and ecosystems, including coral reefs. Another source of pollution comes from harmful algal blooms,[4] which ends up smothering the reef and depleting the water of oxygen. The Save Our Shores Act was introduced to reauthorize the Harmful Algal Bloom and Hypoxia Research and Control Act of 1998 to mitigate these events and their effects. The Coastal Pollution Reduction Act of 1997 was passed to relocate the outfall of a waste treatment facility in Mayaguez Bay, Puerto Rico to avoid polluting the already degraded coral reef. There have been two legislative efforts to protect the Florida Reef Tract by prohibiting offshore oil and gas development, which pollute coral reef ecosystems with oil spills and litter, however these were not successful.

4 Harmful Algal Blooms (HABs) occur when nutrient overload leads to an exponential increase in phytoplankton populations that become toxic to marine organisms in large quantities. They can lead to large mortalities in fish, marine mammal and bird populations where they occur. More specifically, when these populations of phytoplankton die, they create a zone of very low oxygen which is harmful for coral organisms (Smayda, 2008).

Table 9.4 Bills that addressed specific stressors for coral reef ecosystems.

Bill Name	*Congress*	*Year*	*Purpose*	*Status*
H.R. 2258 Hurricane Research Initiative Act of 2011	112	2011–2012	To establish the National Hurricane Research Initiative to improve hurricane preparedness, and for other purposes.	Introduced
H.R. 327 National Hurricane Research Initiative Act of 2009	111	2009–2010	To establish the National Hurricane Research Initiative to improve hurricane preparedness, and for other purposes.	Introduced
H.R. 3888 Clean Cruise Ship Act Of 2009	111	2009–2010	To amend the Federal Water Pollution Control Act to establish national standards for discharges from cruise vessels.	Introduced
H.R. 2407 National Hurricane Research Initiative Act of 2007	110	2007–2008	To establish the National Hurricane Research Initiative to improve hurricane preparedness, and for other purposes.	Introduced
H.R. 1091 Save Our Shores Act	110	2007–2008	Amends the Harmful Algal Bloom and Hypoxia Research and Control Act of 1998 to extend the authorization of appropriations for research, education and monitoring activities related to the prevention, reduction and control of harmful algal blooms and hypoxia.	Introduced
H.R. 1590 Safe Climate Act of 2007	110	2007–2008	To reduce greenhouse gas emissions and protect the climate.	Introduced
H.R. 4174 FOARAM Act	110	2007–2008	To establish an interagency committee to develop an ocean acidification research and monitoring plan and to establish an ocean acidification program within the National Oceanic and Atmospheric Administration.	Passed House
H.R. 6434 Clean Cruise Ship Act of 2008	110	2007–2008	To amend the Federal Water Pollution Control Act to establish national standards for discharges from cruise vessels.	Introduced
H.R. 5642 Safe Climate Act of 2006	109	2005–2006	To reduce greenhouse gas emissions and protect the climate.	Introduced
H.R. 5946 Magnuson-Stevens Fishery Conservation and Management Reauthorization Act of 2006	109	2005–2006	To amend the Magnuson-Stevens Fishery Conservation and Management Act to authorize activities to promote improved monitoring and compliance for high seas fisheries, or fisheries governed by international fishery management agreements, and for other purposes.	Became Law

(Continued)

Table 9.4 (Continued)

Bill Name	*Congress*	*Year*	*Purpose*	*Status*
H.R. 1636 Clean Cruise Ship Act of 2005	109	2005–2006	To amend the Federal Water Pollution Control Act to establish national standards for discharges from cruise vessels.	Introduced
H.R. 5051 Magnuson-Stevens Fishery Conservation and Management Amendments Act of 2006	109	2005–2006	To authorize appropriations to the Secretary of Commerce for the Magnuson-Stevens Fishery Conservation and Management Act for fiscal years 2007 through 2012, and for other purposes.	Introduced
H.R. 1636 Clean Cruise Ship Act of 2005	109	2005–2006	To amend the Federal Water Pollution Control Act to establish national standards for discharges from cruise vessels.	Introduced
H.R. 3778 Bottom Trawl and Deep Sea Coral Habitat Act	109	2005–2006	To establish ocean bottom trawl areas in which trawling is permitted, to protect deep sea corals and sponges, and for other purposes.	Introduced
H.R. 4101 Clean Cruise Ship Act of 2004	108	2003–2004	To amend the Federal Water Pollution Control Act to establish national standards for discharges from cruise vessels.	Introduced
H.R. 2207 Coastal Pollution Reduction Act Of 1997	105	1997–1998	To amend the Federal Water Pollution Control Act concerning a proposal to construct a deep ocean outfall off the coast of Mayaguez, Puerto Rico.	Passed House
H.R. 73 To Protect the Ecologically Fragile Coastal Resources of South Florida	104	1995–1996	To protect the ecologically fragile coastal resources of South Florida by prohibiting offshore oil and gas activities and by canceling Federal leases in the area of the Outer Continental Shelf adjacent to the South Florida coast.	Introduced
H.R. 4613 To Protect the Ecologically Fragile Coastal Resources of South Florida	103	1993–1994	To protect the ecologically fragile coastal resources of South Florida by prohibiting offshore oil and gas activities and by canceling Federal leases in the area of the Outer Continental Shelf adjacent to the South Florida coast.	Introduced

The Magnuson-Stevens Fishery Conservation and Management Reauthorization Act of 2006 and Magnuson-Stevens Fishery Conservation and Management Amendments Act of 2006 both outlined fishery management policy. The Bottom Trawl and Deep Sea Coral Habitat Act specifically prohibits the use of specific fishing gear and fishing in particular deep sea coral habitat in U.S. waters.

Coral reefs are additionally at risk from increasing sea surface temperatures and ocean acidification from global climate change. The Safe Climate Act of 2007 and 2006 amended the Clean Air Act to direct the Environmental Protection Agency to create emission targets for a 2 percent annual reduction in greenhouse gas emissions from 2010 to 2050 and regulations requiring reductions to meet these targets. The Federal Ocean Acidification Research And Monitoring (FOARAM) Act would have required the development of a plan for federal ocean acidification research and monitoring for adaptation and mitigation strategies to conserve marine organisms and ecosystems.

Who Supports Bipartisan Coral Reef Policy?

With partisan gridlock in Congress, it is becoming increasingly important for environmental policies to have bipartisan support from both major political parties to be passed through Congress and become laws (Fisher & Leifeld, 2019). In this section, we present our findings examining the history of which legislators from the House of Representatives have supported the bipartisan bills outlined in the previous section through co-sponsorship, roll-call voting and both actions.

Sponsors of specific policies are legislators who introduce the bills and co-sponsors are other legislators who tie their names to the bill to publicly support it (American Legion, n.d.). This allows us to examine who was willing to publicly support coral reef conservation policies. Roll-call voting is a vote on the floor of Congress to pass a bill that records the vote of each legislator; there are also voice votes, which do not record what stance each legislator took (U.S. Senate, 2021). We also studied which legislators co-sponsored and voted for these policies, to get an understanding of who was most likely to always support coral reef conservation policies in the House of Representatives. This allows us to see which states are least likely to vote for these policies. Our expectations were that Democratic leaning states, or "Blue States," and states with coral reefs and coastlines were more likely to co-sponsor and vote for coral reef conservation policy while Republican leaning states, or "Red States," and land-locked states were not. These findings can be used by advocacy groups and scientists who wish to advocate for similar conservation policy.

Differences between state support for bipartisan coral reef policies

This section analyzes how each state has supported coral reef conservation by examining the number of representatives that (1) co-sponsored, (2) voted for[5] or (3) both simultaneously (e.g., co-sponsored *and* voted for bipartisan coral reef legislation). We standardized the number of representatives for each state to calculate how support varied by each state according to the total number of representatives that each state has ever had. Without standardizing, the most populous states would have the highest frequencies of votes due to raw numbers because of how the House of Representatives delegate counts are determined by population. Standardization helped to address vacancies between election cycles in the House of Representatives.

First, for co-sponsorship, representatives from Hawaii were the most likely to sponsor or co-sponsor bipartisan coral reef legislation, followed by Alaska and Florida (Figure 9.1). Hawaii and Florida are the two U.S. states with adjacent coral reefs. In Hawaii, coral reefs provide ecosystem services valued at $863 million per year, underpinning the importance placed on these reefs by Hawaii's lawmakers (State of Hawaii, 2019). It is estimated that coral reefs contribute approximately $8.5 billion to Florida's local economy annually (Florida Keys National Marine Sanctuary, 2011).

Somewhat perplexing was Alaskan support for coral reef conservation policy. Alaska does not have coral reefs, but it has the longest coastline of any U.S. state, which suggests its rationale. Alaska does not have tropical coral reefs, but it does have widespread deep-sea coral reefs with high levels of biodiversity and the most abundant, deep-water non-reef building corals in the world (NOAA Fisheries, n.d.). Likewise, Alaska representative Don Young has served in the House of Representatives for the entire 32-year period in which coral reef policies have been introduced. He is prominent on issues involving ocean conservation, and currently serves as the co-chair of the House Oceans Caucus, a bipartisan cohort of members of the House of Representatives who prioritize policies related to the health of the ocean. Although Representative Young has not made any public statements on coral reefs since 1997, he submitted a report in support of a concurrent resolution[6] (H. Con. Res. 8) "expressing the sense of Congress with respect to the significance of maintaining the health and stability of coral reef ecosystems"

---

5 If a legislator votes yes on legislation, this means they want it to pass and become a law.

6 A concurrent resolution is a bill in the House of Representatives and Senate that has no law-making powers, but instead makes a statement about what Congress cares about (U.S. Capitol, n.d.).

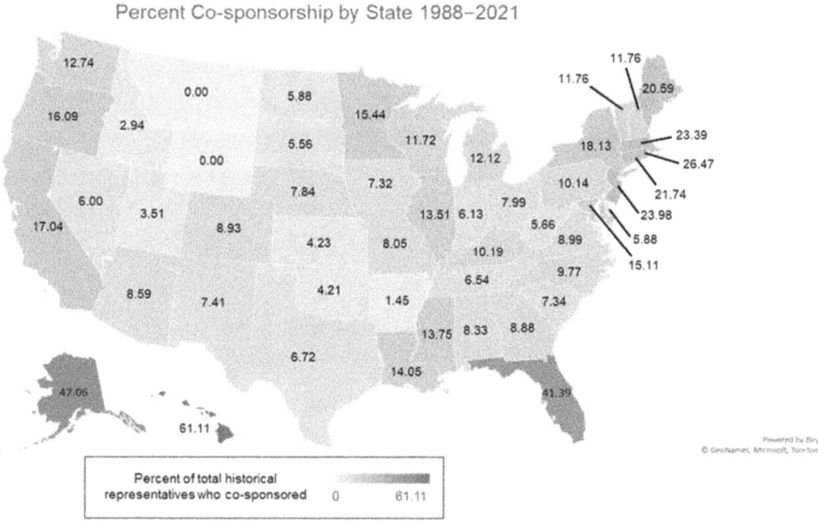

Figure 9.1 Percent co-sponsorship by state for bipartisan coral reef legislation.

(House of Representatives Report No. 105-69, 1997). He has also had a long track record of supporting ocean conservation policies.[7] Representatives from Montana, Idaho, Wyoming and Kansas have never co-sponsored and voted for coral reef conservation legislation. Montana, Idaho, Wyoming and Kansas are landlocked states, meaning they have no coasts or ties to coastal states other than political party and ideology.

Second, for voting for coral reef legislation in Congress, representatives from New Mexico were most likely to vote for bipartisan coral reef legislation, followed by Maine, Vermont, New Hampshire and Delaware (Figure 9.2). All of these states have a large number of representatives who have historically been members of the Democratic Party and have typically voted for the Democratic nominee in the past five presidential elections (Electoral Ventures LLC, n.d.; Monkovic, 2016). During this time period, according to our data, 65 percent of legislators from New Mexico were Democrats, 80 percent of legislators from Maine were Democrats, 66 percent of legislators from Vermont were Democrats, 42 percent of legislators from New Hampshire were Democrats, and 75 percent of legislators from Delaware

7 Representative Young co-sponsored the Coastal and Great Lakes Communities Enhancement Act, Magnuson-Stevens Fishery Conservation and Management Reauthorization Act of 2006, Conservation and Reinvestment Act, Coastal Pollution Reduction Act of 1997 and the Oceans Act of 1992.

Figure 9.2 Percent roll-call votes by state for bipartisan coral reef legislation.

were Democrats. Representatives from Montana and Wyoming never co-sponsored bipartisan coral reef legislation. Montana and Wyoming are land-locked states, meaning they have no coasts or ties to coastal states other than political party and ideology.

Third, for both simultaneously (where lawmakers both co-sponsor *and* vote for legislation), representatives from Michigan and Alaska had the highest number of past co-sponsors and votes for bills supporting coral reef conservation, followed by states with coral reefs, Hawaii and Florida (Figure 9.3). Again, Florida and Hawaii are the two states with coral reefs. Michigan has a political reputation for being a Democratic-leaning "blue wall state,"[8] a trend which is supported by the states' past legislators voting and co-sponsoring the following coral reef conservation bills: Coastal and Great Lakes Communities Enhancement Act, Coral Reef and Coastal Marine Conservation Act of 2006, Conservation and Reinvestment Act and Coastal Zone Act Reauthorization Amendments of 1990.

8 A blue wall state is the collection of Midwestern states, including Michigan, that have historically been important for supporting Democratic nominees for president and securing Democratic votes in the electoral college for U.S. presidential elections (Agnew & Shin, 2021).

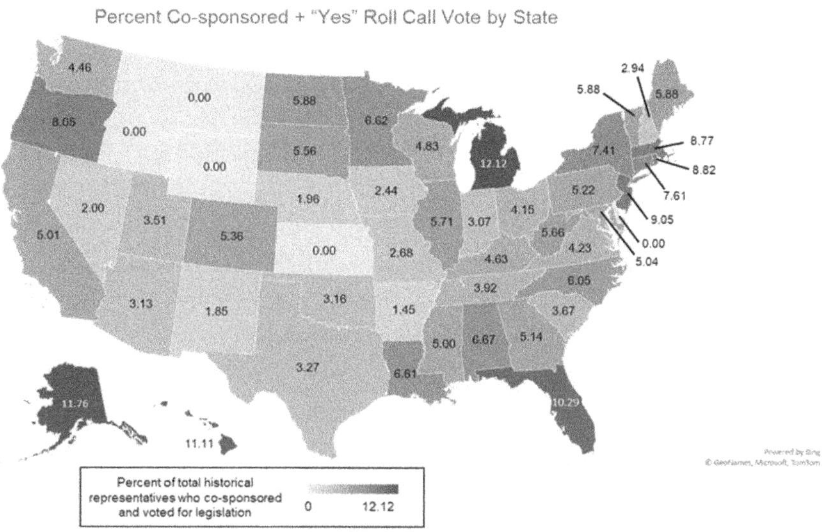

Figure 9.3 Percent co-sponsorship and roll-call votes for bipartisan coral reef legislation.

Are Coral Reefs a Popular Symbol That Can Be Used to Encourage Support for Climate Policy?

Coral reefs are climate-vulnerable ecosystems. There is a historical tendency in the Republican Party to deny the legitimacy of climate science, refute the expertise of specific climate scientists and use the processes of Congress to keep climate change related policies off the voting floor of Congress (McCright & Dunlap, 2010). Since the 1980s, the partisan gap over climate policy has widened over each year examined in the House of Representatives (Gershtenson et al., 2006). Even when other factors (e.g., constituency demographics) were considered, political party was still one of the most important reasons congressional members in either chamber voted a certain way on environmental legislation (with Democrats voting yes and Republicans voting no) (Gershtenson et al., 2006).

Guber et al. (2021) found that Democrats and Republicans in Congress were significantly different in their messaging on climate change, where Democrats focused on evidence-based messages on the impacts to human society and Republicans focused on anecdotal denial of the threats from climate change. Polarization has led to gridlock characterized by a lack of policy-making on climate change. Gridlock occurs in multiple levels of government, with groups of legislators from the U.S. Congress and state legislatures working together to delay climate policy (Fisher & Leifeld, 2019).

There is some evidence that climate may become a bipartisan issue. Potential reasons are increasing risk perceptions (Heazle et al., 2013; O'Connor et al., 1999) and increased media coverage of related hazards (Carmichael & Brulle, 2016; Liu et al., 2011; Mccright & Dunlap, 2011). Previous opponents of climate change policy in the Republican Party may be facing pressure from the increase in voters who want to see their representatives address these issues (Krosnick & MacInnis, 2020). Republican legislators may be open to compromise with older constituents who are anti-climate policy and newer constituents who want to see climate acknowledged by their policy-makers (Van Boven et al., 2018). Support for climate change may change when lawmakers experience impacts first hand. Gagliarducci et al. (2019) show that when natural disasters associated with climate change increase, co-sponsorship and voting for climate policies may increase.

An example of bipartisan coral reef conservation policy that is linked to climate change is the Restoring Resilient Reefs Act of 2021. While the text of the Restoring Resilient Reefs Act does not explicitly say the words, "climate change," it does mention that coral reef ecosystems are facing both natural and "human-accelerated changes," including increasing ocean temperatures, ocean acidification and stressors like coral bleaching which are attributed to climate change (Restoring Resilient Reefs Act, 2019).

The Restoring Resilient Reefs Act legislation is important because it demonstrates the most up-to-date thinking in Congress for creating coral reef conservation policy in the United States. The Restoring Resilient Reefs Act aims to reauthorize[9] an important law aimed at coral reef conservation titled the Coral Reef Conservation Act of 2000,[10],[11] and is the latest version of this

9 Reauthorization is a necessary legislative action for approving and funding activities in federal agencies in the executive branch that enact the laws passed by Congress (Heniff, 2012).

10 The Coral Reef Conservation Act of 2000 expired in 2004 and there have been multiple attempts to introduce reauthorization bills in both chambers almost annually since then. The programs that The Coral Reef Conservation Act created remain in place through appropriations made each year in Congress, due to the existence of a system where some U.S. laws can fail to be reauthorized, but their funding can remain in place through the original legislation due to continuing appropriations. Appropriations in American lawmaking occur when a law provides a specific budget to a federal agency to perform the duties described in the policy (United States House of Representatives, n.d.). In this case, appropriations allowed for the continued funding of coral reef conservation activities in the United States, even if Congress did not reauthorize the law. A summary of the bills to reauthorize the Coral Reef Conservation Act of 2000 is in Table 9.6.

11 The result of President Clinton's Executive Order No. 13089.

law (table 9.5 provides an overview of the history of this law). This earlier version of the Restoring Resilient Reefs Act was the first time that dedicated funding and protection for coral reefs was written into U.S. law to support federal management of coral reef ecosystems and create the U.S. Coral Reef Initiative, a program in NOAA to promote federal protection for the nation's coral reefs. The Restoring Resilient Reefs Act legislation shows that some language around a changing climate is permitted in bills supported by both parties. The Restoring Resilient Reefs Act is still in the process of being passed in both chambers of Congress and becoming a law; it has been introduced three times in the past three congressional sessions.

The main purpose of the Restoring Resilient Reefs Act is to provide resources and a federal government-facilitated means to collect private donations to fund coral reef conservation projects. The Restoring Resilient Reefs Act[12] has a unique structure that would increase spending on coral reef conservation over time by authorizing increased appropriations until the expiration of the law in the next congressional session. Funds are to be used for federal coral reef management programs, specifically for activities such as monitoring and education in federal MPAs and state grants to support community-based management programs.

Community-based management programs are when local people that depend on reefs are invited by leaders to be actively involved in the management of their resources (Ostrom, 1990). Federal coral reef management is increasingly tapping into this style of project to leverage local community knowledge and buy-in to engage in reef conservation. An example is the *Mission: Iconic Reefs* program which is made up of U.S. agencies, non-profits and scientific teams to restore and conserve the Florida Reef Tract (NOAA, 2021). This program is a strategic plan for conducting studying biological, climatic and socioeconomic stressors impacting reefs in U.S. states and territories. The resulting data provide a robust picture of the condition of U.S. coral reef ecosystems and the communities connected to them, and where interventions can be made for conservation. Additional funds from this policy are allocated for research, state agency-led conservation programs, emergencies such as responding to disease outbreaks (Restoring Resilient Reefs Act, 2019).

The Restoring Resilient Reefs Act modernizes the U.S. Coral Reef Task Force, one of the most important decision-making organizations in the U.S. for coral reef management. The U.S. Coral Reef Task Force is a government body with responsibilities to conserve, restore and monitor coral reef

12 Section 224 of the bill specifically.

Table 9.5 History of efforts to reauthorize the Coral Reef Conservation Act of 2000.

Name of Bill	*Congress*	*Bipartisan (D=Democratic led, R=Republican led)*	*Senate*	*House*	*Reached Vote?*
Restoring Resilient Reefs Act of 2021	117th	Yes	No	Yes	No
Coral Reef Conservation Reauthorization Act of 2020	116th	Yes	Yes	Yes	No
The Restoring Resilient Reefs Act of 2019 and the Restoring Resilient Reefs Act of 2020	116th	Yes	Yes	Yes	Passed Senate
Coral Reef Sustainability Through Innovation Act of 2019	116th	Yes	Yes	Yes	No
Coral Reef Conservation Reauthorization Act of 2018	115th	Yes	No	Yes	No
Coral Reef Sustainability Through Innovation Act of 2017	115th	D	Yes	Yes	No
Coral Reef Sustainability Through Innovation Act of 2016 (2)	114th	D	Yes	Yes	No
Coral Reef Conservation Act Reauthorization and Enhancement Amendments of 2016	114th	D	No	Yes	No
Conserving Our Reefs and Livelihoods Act	114th	R	No	Yes	No
Coral Reef Conservation Amendments Act of 2013	113th	D	Yes	No	No
Coral Reef Conservation Act Reauthorization and Enhancement Amendments of 2013	113th	D	No	Yes	No
Coral Reef Conservation Amendments Act of 2011	112th	D	Yes	No	No
Coral Reef Conservation Act Reauthorization and Enhancement Amendments of 2011	112th	D	No	Yes	No
Coral Reef Conservation Amendments Act of 2009	111th	D	Yes	No	No
Coral Reef Conservation Act Reauthorization and Enhancement Amendments of 2009	111th	D	No	Yes	Passed House
Coral Reef Ecosystem Conservation Amendments Act of 2007	110th	D	Yes	No	No
Coral Reef Conservation Amendments Act of 2007	110th	Yes	Yes	Yes	Passed House
Coral Reef Conservation Amendments Act of 2006	109th	D	No	Yes	No
Coral Reef Conservation Legacy Act of 2006	109th	R	No	Yes	No
Coral Reef Conservation Amendments Act of 2005	109th	D	Yes	No	Passed Senate
Coral Reef Conservation and Protection Act of 2004	108th	D	No	Yes	No
Coral Reef Conservation Act of 2000		Executive Order 13089			

ecosystems located within federal natural resource agencies (Craig, 2000). The U.S. Coral Reef Task Force, established with the passage of the Coral Reef Conservation Act of 2000, has expanded in members and responsibilities since its early days. Agencies with roles in the U.S. Coral Reef Task Force include NOAA, the National Park Service,[13] U.S. Fish and Wildlife Service, the U.S. Geological Survey (USGS), and The Office of Insular Affairs[14] (*U.S. Coral Reef Task Force Members*). The Restoring Resilient Reefs Act legislation specifically grants them authority to implement policy and management strategies when the legislation is passed (Restoring Resilient Reefs Act, 2019). Decisions are made through voting, with votes held by representatives from the federal agencies and U.S. states, but with U.S. territories having non-voting memberships to observe and offering insight[15] (Craig, 2000).

The Restoring Resilient Reefs Act was first introduced in the House of Representatives by Democrat Darren Soto from Florida with 22 co-sponsors[16] in 2019. It was then introduced in the Senate by Republican (R) Marco Rubio, with three bipartisan co-sponsors: Brian Schatz, a Democrat (D), Mazie K. Hirono (D) and Rick Scott (R). Senator Rubio's sponsorship is of note because he frequently criticized federal government spending. His

---

13 Which is housed in the U.S. Department of the Interior (as is U.S. Fish and Wildlife, the USGS and the Office of Insular Affairs).

14 The U.S. Office of Insular Affairs oversees the administration of various U.S. island areas, including all U.S. Island Territories. Other Task Force agencies include the Mineral Management Service, Natural Resource Conservation Service, U.S. Agency for International Development, U.S. Coast Guard, U.S. Department of Agriculture Field Office, U.S. Navy, U.S. Army Corps of Engineers, U.S. Department of State, U.S. Department of Transportation, Environmental Protection Agency, National Science Foundation, and National Aeronautics and Space Administration.

15 State and Territory voting members are the Commonwealth of the Northern Mariana Islands, Commonwealth of Puerto Rico, State of Florida, State of Hawaii, Territory of Guam, Territory of American Samoa, Territory of the U.S. Virgin Islands and non-voting members include the Federated States of Micronesia, Republic of Marshall Islands and Republic of Palau.

16 Resident Commissioner González-Colón, Jenniffer [R-PR-At Large], Rep. Crist, Charlie [D-FL-13], Rep. Mast, Brian J. [R-FL-18], Rep. Murphy, Stephanie N. [D-FL-7], Del. Radewagen, Aumua Amata Coleman [R-AS-At Large], Rep. Gabbard, Tulsi [D-HI-2], Rep. Fitzpatrick, Brian K. [R-PA-1], Rep. Rush, Bobby L. [D-IL-1], Rep. Deutch, Theodore E. [D-FL-22], Rep. Lawson, Al, Jr. [D-FL-5], Del. Sablan, Gregorio Kilili Camacho [D-MP-At Large], Rep. Posey, Bill [R-FL-8], Rep. Frankel, Lois [D-FL-21], Rep. Rooney, Francis [R-FL-19], Rep. Waltz, Michael [R-FL-6], Rep. Yoho, Ted S. [R-FL-3], Rep. Mucarsel-Powell, Debbie [D-FL-26], Rep. Buchanan, Vern [R-FL-16],Rep. Castor, Kathy [D-FL-14], Rep. Bilirakis, Gus M. [R-FL-12], Rep. Gaetz, Matt [R-FL-1], Rep. Webster, Daniel [R-FL-11] (Restoring Resilient Reefs Act, 2019).

Table 9.6 Legislators who sponsored* and co-sponsored the Restoring Resilient Reefs Act of 2021 who were included in our data collection.

Name	*State*	*Political Party*	*Congressional Chamber*
Darren Soto*	Florida	Democratic	House of Rep.
Marco Rubio*	Florida	Republican	Senate
Rick Scott	Florida	Republican	Senate
Brian Mast	Florida	Republican	House of Rep.
Ed Case	Hawaii	Democratic	House of Rep.
Brian Schatz	Hawaii	Democratic	Senate
Mazie Hirono	Hawaii	Democratic	Senate
Stacey E. Plaskett	US Virgin Islands	Republican	House of Rep.
Jenniffer Gonzalez-Colon	Puerto Rico	Republican	House of Rep.
Aumua Amata Coleman Radewagen	American Samoa	Republican	House of Rep.

shifting attitude on federal spending regarding reefs received widespread praise from local politicians in his home state of Florida, from scientists and environmental non-profits (Hudson, 2019). The original sponsors and co-sponsors whose statements were collected for our dataset are in Table 9.6

Rationale for coral conservation policy support among lawmakers: Community-based conservation with rapid response potential

These final two sections examine policy-maker rationale for supporting coral reef conservation using 253 quotes from lawmakers themselves sources using methods outlined in Appendix for Chapter 8. Most lawmakers discussed the concept of *governance*, or the various management organizations tasked with decision-making and implementing conservation programs. There was widespread agreement among lawmakers that the federal government should be involved with reef conservation but that it should be in a role that supports communities and local actors to be the main actors in conservation projects. There is also agreement over the importance of bicameral, bipartisan public policy-making and the need for urgent responses to stressors like coral disease. The topic of Governance did not have a mathematically significant difference in its use between Democrats and Republicans (Democrats 16.7 percent of messaging, Republicans 15.9 percent of messaging, p-value 0.34).

The first topic of governance focused on the political processes of enacting and implementing coral conservation legislation. This included federal agencies' authority to protect reefs and implement policies. For example, Republican Florida Senator Marco Rubio, when announcing the Restoring

Resilient Reefs Act argues that it is a "comprehensive response [...] to ensure federal agencies are partnering effectively with state and local governments, as well as the communities who rely on the vitality of these critical habitats" (Targeted News Service, 2020). Two components of this excerpt are important, first being that federal agencies must work together to protect reefs as efficiently as possible, through relying on state and local governments to implement some aspects of conservation. Second, that local communities will be important partners in implementing reef conservation.

Republican Congresswomen Gonzalez-Colon from Puerto Rico echoes these sentiments where state and local partners play an important role in federal conservation by highlighting how this bill allows:

> *NOAA, to provide state grants to allow U.S. coastal coral jurisdictions, like Puerto Rico, to apply for funding to meet the priorities they have set for restoration and management of their own coral reef.* (House Natural Resources Subcommittee on Water, Oceans and Wildlife Hearing; The Transformation of the Puerto Rico Electric Power Authority (PREPA), 2020)

Speed is another component of the Restoring Resilient Reefs Act of 2021 as it contains emergency funds to address coral reef emergencies, like disease outbreaks, invasive species, coral bleaching, natural disasters, among many others, with an emphasis on rapid response.

Other statements involving governance focused on the importance of bipartisan congressional lawmaking in Congress where lawmakers said that this support is key to reauthorizing existing federal programs and "continuing the desperately needed programs halting the deterioration of coral reef[s]," (U.S. Representative Darren Soto, 2021). This is noteworthy, as it contradicts conventional wisdom in the literature, where legislators have claimed that "bipartisan deals are dangerous, more often than not" because "pressures from the [party] bases get in the way" (Lee, 2016, p. 50). Lee also describes how polarization and party politics in Congress that lead to bipartisan cooperation are seen as a weak point for the majority party and a power play by the minority party. However, the majority party would be more likely to cooperate to successfully pass legislation, even at the cost of agreeing with the opposing party which reflects badly on their reputation (Lee, 2016).

Another focus of statements on governance by lawmakers was on the importance of community-based institutions for conservation versus more centralized management. According to a statement by Democratic senator Brian Schatz from Hawaii, "federal resources to the local governments and community organizations that are in the water right now working to restore our reefs" are needed for funding community-based management for reefs in the United States (United States Senator Marco Rubio, 2019). He also emphasized the importance of "participatory community-based management" which unlocks the creativity and interest of stakeholders who care to be involved with reef management (United States Senator Brian Schatz, 2019).

Democratic senator Mazie Hirono of Hawaii highlights the significant amount of attention paid to empowering local-scale institutions to manage coral reefs at the community level, signaling community-based coral reef management as a policy priority for Congress. She says, "the resources provided by this bill are urgently needed by local managers as threats to corals continue to increase" (United States Senator Marco Rubio, 2019). Florida Republican Congressman Brian Mast also emphasized that "this legislation would help communities better manage coral reef ecosystems and protect our coastlines" (U.S. Representative Brian Mast, 2021). This suggests that the bipartisan vision of the future of coral reef conservation policy in the United States is focused on community-based coral reef management. This follows the contemporary global trend of coral reef management shifting away from centralized government control and into the hands of local communities and organizations for their management (Christie et al., 1994; Cinner et al., 2009, 2012; White & Vogt, 2000).

Rationale for coral conservation policy support among lawmakers: Symbolism

It is possible that lawmakers agree on the importance of coral reef conservation because reefs are political symbols. Symbols are ideas with an emotional impact that allow people to highlight one dimension of a problem while downplaying others (in our research, we construe reefs as a symbol which allows policy-makers to emphasize popular actions like conservation and downplay contentious ideas like climate change) (Zahariadis, 2007). Symbols are related to whether a policy has feasibility, or is perceived as capable of addressing a problem (Pralle, 2009). Our research argues that coral reef conservation (conservation of a climate vulnerable ecosystem) is the pathway to addressing climate impacts that lawmakers perceive as feasible (facilitated through the use of symbols).

Coral reefs were indeed used by lawmakers in our data as symbols, or used to highlight one dimension of a problem (such as coral mortality) while downplaying others (such as the plausible cause of mortality being from warming seas) (Stone, 2011). The symbolic nature of coral reefs, and their popularity, are an ideal point of focus for lawmakers who, because of partisan politics, may be unable to speak directly about climate change and may instead opt to focus on climate vulnerable ecosystems.

Most speakers who employed the theme of symbols did so in order to use coral reefs as a symbol of a particular U.S. state by mentioning local places of high cultural importance and meaning. For example, Republican Florida Senator Marco Rubio provides a description of the Florida Reef Tract as "an integral component of the economic and ecological character of Florida" (United States Senator Marco Rubio, 2019). Rubio's statements were representative, saying that reefs are, in effect, synonymous with the state of Florida. This is referencing the importance of the diving and snorkeling tourism in Florida, which has been estimated to generate billions of dollars annually from thousands of visitors a year (Wynveen et al., 2013).

Democratic representative Ed Case from Hawaii highlighted the state's "precious reefs stretch 1,500 miles across the blue Pacific from Hawai'i Island to Midway Atoll" (U.S. Representative Darren Soto, 2020). Midway Atoll is an iconic and important historical landmark from one of the most decisive battles of World War II (U.S. Fish & Wildlife Service, n.d.). Republican Congresswoman Amata from American Samoa recognizes that the Restoring Resilient Reefs Act "recognizes the importance of [our] reefs, which include some of the oldest and largest reefs in the world, and supports their preservation" (Targeted News Service, 2021).

Democratic Florida representative Darren Soto and Democratic Hawaii senator Mazie Hirono described coral reefs as essential to the "way of life" in Florida and Hawaii (Targeted News Service, 2020; U.S. Representative Darren Soto, 2021). This sentiment is echoed by Republican Florida Senator Marco Rubio describing coral reefs as "characters of Florida" that are integral to representing the state's culture (Targeted News Service, 2020).

Other lawmakers saw reefs as a symbol of the obligation that we have to pass on nature to future generations, such as Republican Florida Senator Rick Scott's statement that the significance of this legislation is to "make sure future generations can enjoy all that Florida has to offer" in terms of coral reef resources (United States Senator Marco Rubio, 2019). The importance of conserving coral reef ecosystems as iconic national symbols to safeguard for future generations was also supported by Republican senator Marco Rubio, Republican senator Rick Scott, Democratic representative Darren Soto and Democratic representative Mazie Hirono (United States Senator Marco Rubio, 2019). Members of non-profits including the Florida Keys Commercial Fishermen's Association, National Marine Manufacturers Association, Aquarium Conservation Partnership, Shedd Aquarium and the decision-makers like the secretary of the Florida Department of Environmental Protection also emphasized this benefit of protecting coral reef ecosystems for future generations.

A Pathway for Policies Regarding Climate-Vulnerable Ecosystems

Our research provides insight on how coral reef conservation policies have been made into the U.S. legislature and how lawmakers are beginning to use coral reefs as symbols because it prevents them from having to mention climate change. Our data shows that coral reef policy created by Congress is becoming more emphasized on local communities and community-based projects to conserve reefs as well as on rapid responses to worsening coral diseases. The local-level project bases and the emphasis on rapid-response capacity suggests that adaptive governance of coral reefs is taking place, similar to that described in Chapter 5 on the FKNMS.

Our findings contribute to a conversation started by Gershtenson and colleagues (2006) who also found that although political party mattered when debating climate change policy, sometimes it was not the most important factor when constituency demographics were taken into account. Understanding what motivates policies for the conservation of climate-vulnerable systems like reefs is key for the future of climate policy. This is especially crucial as more lawmakers of both political parties are supporting these kinds of policies in ways not observed with more polarizing policies aimed at climate like the Green New Deal.

References

Ackrill, R., Kay, A., & Zahariadis, N. (2013). *Ambiguity, multiple streams, and EU policy.* https://doi.org/10.1080/13501763.2013.781824

Agnew, J., & Shin, M. (2021). The fragile blue wall: Analyzing geographies of the 2020 US presidential election. *Quaderni Dell'Osservatorio Elettorale – Italian Journal of Electoral Studies, 84*(1), 31–44. https://doi.org/10.36253/qoe-10161

Alagiri, P. (1992). Give us sovereignty or give us debt: Debtor countries' perspective on debt-for-nature swaps. *American University Law Review, 41*(2), 485–515.

Bergquist, P. (2020). Congress as theatre: How advocates use ambiguity for political advantage. *Journal of Public Policy, 40*, 51–71. https://doi.org/10.1017/S0143814X18000284

Brunner, S. (2008). Understanding policy change: Multiple streams and emissions trading in Germany. *Global Environmental Change, 18*(3), 501–507. https://doi.org/10.1016/j.gloenvcha.2008.05.003

Cain, B. E., Gerber, E. R., & Hui, I. (2020). Getting bipartisan support for sea level rise adaptation policies. *Ocean and Coastal Management, 197*, 105298. https://doi.org/10.1016/j.ocecoaman.2020.105298

Carmichael, J. T., & Brulle, R. J. (2016). Elite cues, media coverage, and public concern: An integrated path analysis of public opinion on climate change, 2001–2013. *Environmental Politics.* https://doi.org/10.1080/09644016.2016.1263433

Christie, P., White, A. T., & Buhat, D. (1994). Community-based coral reef management on San Salvador Island, the Philippines. *Society and Natural Resources, 7*(2), 103–117. https://doi.org/10.1080/08941929409380849

Cinner, J. E., McClanahan, T. R., Daw, T. M., Graham, N. A., Maina, J., Wilson, S. K., et al. (2009). Linking social and ecological systems to sustain coral reef fisheries. *Current Biology*, *19*, 206–212. https://doi.org/10.1016/j.cub.2008.11.055

Cinner, J. E., McClanahan, T. R., MacNeil, M. A., Graham, N. A., Daw, T. M., Mukminin, A., Feary, D. A., Rabearisoa, A. L., Wamukota, A., Jiddawi, N., & Campbell, S. J. (2012). Comanagement of coral reef social-ecological systems. *Proceedings of the National Academy of Sciences*, *109*(14), 5219–5222.

Craig, R. K. (2000). The coral reef task force: Protecting the environment through executive order. *Environmental Law Reporter*, *30*, 10343–10364. http://ssrn.com/abstract=922408

Dolan, D. A. (2021). Multiple partial couplings in the multiple streams framework: the case of extreme weather and climate change adaptation. *Policy Studies Journal*, *49*(1), 164–189. https://doi.org/10.1111/PSJ.12341

Electoral Ventures LLC. (n.d.). Blue and Red States. https://www.270towin.com/content/blue-and-red-states

Fisher, D. R., & Leifeld, P. (2019). The polycentricity of climate policy blockage. *Climatic Change*, *155*(4), 469–487. https://doi.org/10.1007/s10584-019-02481-y

Florida Keys National Marine Sanctuary. (2011, April 4). Coral reefs support jobs, tourism, and fisheries. https://floridakeys.noaa.gov/corals/economy.html

Floor, J. R., Van Koppen, K., & Van Tatenhove, J. P. M. (2018). *Science, uncertainty and changing storylines in nature restoration: The case of seagrass restoration in the Dutch Wadden Sea.* https://doi.org/10.1016/j.ocecoaman.2018.02.016

Gagliarducci, S., Paserman, D. M., & Patacchini, E. (2019). Hurricanes, climate change policies, and electoral accountability. In *National Bureau of Economic Research Working Paper Series.* https://doi.org/10.3386/w25835

Garcia Hernandez, A. L., & Bolwig, S. (2020). Understanding climate policy integration in the global South through the multiple streams framework. *Climate and Development*, *13*(1), 68–80. https://doi.org/10.1080/17565529.2020.1723471

Gershtenson, J., Smith, B. W., & Mangun, W. R. (2006). Friends of the earth? Partisanship, party control of congress, and environmental legislation in congress. *Politics and Policy*, *34*(1), 66–92. https://doi.org/10.1111/j.1747-1346.2006.00004.x

Goncalves, L., & De Santo, E. M. (2021). Unpacking the process: How agenda-setting theory explains the case of creating large scale marine protected areas in Brazil. *Environmental Politics*, 1–20. https://doi.org/10.1080/09644016.2021.1915037

Goyal, N. (2021). Policy diffusion through multiple streams: The (non-)adoption of energy conservation building code in India. *Policy Studies Journal*, 1. https://doi.org/10.1111/PSJ.12415

Guber, D. L., Bohr, J., & Dunlap, R. E. 2021. 'Time to wake up': climate change advocacy in a polarized Congress, 1996–2015. *Environmental Politics*, *30*(4), 538–558.

Harbridge, L. M. (2009). *Bipartisanship in a polarized congress.* Stanford University.

Heazle, M., Tangney, P., Burton, P., Howes, M., Grant-Smith, D., Reis, K., & Bosomworth, K. (2013). Mainstreaming climate change adaptation: An incremental approach to disaster risk management in Australia. *Environmental Science*, *33*, 162–170. https://doi.org/10.1016/j.envsci.2013.05.009

Heniff, B. (2012). Overview of the authorization-appropriations process. www.crs.gov

House concurrent resolution 8. House Report 105-69, Date: April 23, 1997. Congressional sessions: 105–1. https://www.congress.gov/105/crpt/hrpt69/CRPT-105hrpt69.pdf

House Natural Resources Subcommittee on Water, Oceans and Wildlife Hearing; The Transformation of the Puerto Rico Electric Power Authority (PREPA), 116th Cong. (2020). (Testimony by Representative Jenniffer Gonzalez-Colon). https://naturalresources.house.gov/hearings/the-transformation-of-the-puerto-rico-electric-power-authority-prepa

Hudson, H. (2019, November 15). Rubio's Bipartisan coral reef legislation receiving widespread support. http://www.southdadenewsleader.com/news/rubio-s-bipartisan-coral-reef-legislation-receiving-widespread-support/article_b859ccfe-0752-11ea-96d6-f3128e87d685.html

Jones, S. (2014). Flirting with climate change: A comparative policy analysis of subnational governments in Canada and Australia. *Journal of Comparative Policy Analysis: Research and Practice*, *16*(5), 424–440. https://doi.org/10.1080/13876988.2014.942570

Krosnick, J. A., & MacInnis, B. (2020). *Climate insights 2020: Overall trends*. Report. https://doi.org/10.1007/978-3-030-29966-8_11

Kustic, A. (2020). Politics stops at nature's edge: The need for bipartisanship in environmental policymaking. *Political Science & International Studies* | Senior Theses, 5. https://doi.org/10.33015/DOMINICAN.EDU/2020.POL.ST.05

Lee, F. E. (2016). *Insecure majorities: Congress and the perpetual campaign*. The University of Chicago Press.

Library of Congress. (n.d.). Committees of the U.S. Congress. Retrieved February 15, 2022, from https://www.congress.gov/committees

Liu, X., Lindquist, E., & Vedlitz, A. (2011). Explaining media and congressional attention to global climate change, 1969–2005: An empirical test of agenda-setting theory. *Political Research Quarterly*, *64*(2), 405–419. https://doi.org/10.1177/1065912909346744

Llamosas, C., Upham, P., & Blanco, G. (2018). Multiple streams, resistance and energy policy change in Paraguay (2004–2014). *Energy Research and Social Science*, *42*, 226–236. https://doi.org/10.1016/j.erss.2018.03.011

Mason, L. (2018). *Uncivil agreement: How politics became our identity*. The University of Chicago Press.

McCright, A. M., & Dunlap, R. E. (2010). Anti-reflexivity: The American conservative movement's success in undermining climate science and policy. *Theory, Culture and Society*, *27*(2), 100–133. https://doi.org/10.1177/0263276409356001

McCright, A. M., & Dunlap, R. E. (2011). The politicization of climate change and polarization in the American public's views of global warming, 2001–2010. *Sociological Quarterly*, *52*(2), 155–194. https://doi.org/10.1111/j.1533-8525.2011.01198.x

McFarland, B. J. (2018). The origins and history of conservation finance. In *Conservation of tropical rainforests* (pp. 121–131). Palgrave Macmillan. https://doi.org/10.1007/978-3-319-63236-0_6

Messham, E., & Sheard, S. (2020). Taking the long view: The development of the Well-being of Future Generations (Wales) Act. *Health Research Policy and Systems*, *18*(33), 1–12. https://doi.org/10.1186/s12961-020-0534-y

Mintrom, M., & Luetjens, J. (2017). Policy entrepreneurs and problem framing: The case of climate change. *Environment and Planning C: Politics and Space*, *35*(8), 1362–1377. https://doi.org/10.1177/2399654417708440

Monkovic, T. (2016, August 22). 50 years of electoral college maps: How the U.S. turned red and blue. *The New York Times*. https://www.nytimes.com/2016/08/23/upshot/50-years-of-electoral-college-maps-how-the-us-turned-red-and-blue.html

NOAA. (2021, April 23). Restoring seven iconic reefs: A mission to recover the coral reefs of the Florida keys. *NOAA Fisheries*. https://www.fisheries.noaa.gov/southeast/habitat-conservation/restoring-seven-iconic-reefs-mission-recover-coral-reefs-florida-keys

NOAA Fisheries. (n.d.). Deep-sea corals and sponge research in Alaska. NOAA. https://www.fisheries.noaa.gov/alaska/habitat-conservation/deep-sea-corals-and-sponge-research-alaska#:~:text=Deep%2Dsea%20corals%20are%20widespread,the%20world%20are%20found%20here

O'Connor, R. E., Bord, R. J., & Fisher, A. (1999). Risk perceptions, general environmental beliefs, and willingness to address climate change. *Risk Analysis*, *19*(3), 461–471. https://doi.org/10.1023/A:1007004813446

Ostrom, E. 1990. *Governing the commons: The evolution of institutions for collective action*. Cambridge University Press.

Patterson, A. (1990). Debt-for-nature swaps and the need for alternatives. *Environment: Science and Policy for Sustainable Development*, *32*(10), 5–32. https://doi.org/10.1080/00139157.1990.9929058

Pralle, S. B. (2009). Agenda-setting and climate change. *Environmental Politics*, *18*(5), 781–799. https://doi.org/10.1080/09644010903157115

Restoring Resilient Reefs Act of 2019, H.R.4160, 116th Cong. (2019). https://www.congress.gov/bill/116th-congress/house-bill/4160

Richmond, R. H., Rongo, T., Golbuu, Y., Victor, S., Idechong, N., Davis, G., Kostka, W., Neth, L., Hamnett, M., & Wolanski, E. (2007). Watersheds and coral reefs: Conservation science, policy, and implementation. *BioScience*, *57*(7), 598–607. https://doi.org/10.1641/B570710

Saldaña, J. (2016). *The coding manual for qualitative researchers* (3rd ed.). SAGE.

Smayda, T. J. (2008). Complexity in the eutrophication-harmful algal bloom relationship, with comment on the importance of grazing. *Harmful Algae*, *8*(1), 140–151. https://doi.org/10.1016/j.hal.2008.08.018

Smith, Z. A. (2004). *The environmental policy paradox*. Prentice Hall.

State of Hawaii. (2019, December 5). Coral reefs. Division of Aquatic Resources. https://dlnr.hawaii.gov/dar/habitat/coral-reefs/

Stone, D. (2011). *Policy paradox: The art of political decision making* (3rd ed.). W. W. Norton & Company.

Storch, S., & Winkel, G. (2013). Coupling climate change and forest policy: A multiple streams analysis of two German case studies. *Forest Policy and Economics*, *36*, 14–26. https://doi.org/10.1016/J.FORPOL.2013.01.009

Targeted News Service. (2020, December 23). Sen. Rubio, colleagues applaud passage of restoring resilient reefs act. Targeted News Service. https://advance.lexis.com/api/document?collection=news&id=urn:contentItem:61KD-YNY1-DYG2-R369-00000-00&context=1516831

Targeted News Service. (2021, February 27). Reps. Chabot, Sherman tropical forest conservation reauthorization passes committee. Targeted News Service. https://advance.lexis.com/api/document?collection=news&id=urn:contentItem:623H-1V51-DYG2-R332-00000-00&context=1516831

Targeted News Service. (2021, May 7). Del. Amata focuses on coral reef legislation. https://advance.lexis.com/api/document?collection=news&id=urn:contentItem:62M6-VWC1-DYG2-R4X0-00000-00&context=1516831

U.S. Capitol. (n.d.). Two bodies, one branch. U.S. Capitol Visitor Center. https://www.visitthecapitol.gov/about-congress/two-bodies-one-branch

U.S. Coral Reef Task Force Members. (2008, May 20). U.S. coral reef task force, national oceanic and atmospheric administration. www.coralreef.gov/about/members.html

U.S. House of Representatives. (n.d.). Glossary of terms: Appropriation. https://www.house.gov/the-house-explained/open-government/statement-of-disbursements/glossary-of-terms

U.S. Senate. (2021, August 9). Votes. https://www.senate.gov/reference/Index/Votes.htm

United States Representative Brian Mast. (2021). Bill introduced to protect coral reefs [Press Release]. https://mast.house.gov/2021/1/bill-introduced-to-protect-coral-reefs

United States Representative Darren Soto. (2021, January 26). Soto leads legislation to restore and conserve America's coral reef [Press Release]. https://soto.house.gov/media/press-releases/soto-leads-legislation-restore-and-conserve-america-s-coral-reef

United States Senator Brian Schatz. (2019, November 14). Schatz Bill to restore and conserve Hawaii's coral reefs passes key senate committee [Press Release]. https://www.schatz.senate.gov/news/press-releases/schatz-bill-to-restore-and-conserve-hawaiis-coral-reefs-passes-key-senate-committee

United States Senator Marco Rubio. (2019, August 2). Rubio, Schatz, Scott, Hirono introduce bill to restore and conserve America's coral reefs [Press Release]. https://www.rubio.senate.gov/public/index.cfm/press-releases?ID=78CBD372-4BE3-4869-8E51-553709C01F7A

Van Boven, L., Ehret, P. J., & Sherman, D. K. (2018). Psychological barriers to bipartisan public support for climate policy. *Perspectives on Psychological Science, 13*(4), 492–507. https://doi.org/10.1177/1745691617748966

Waldeck, D. A., & Buck, E. H. (2001). The Magnuson-Stevens Fishery Conservation and Management Act: Reauthorization issues for the 107th Congress. In Congressional Research Service Report.

White, A. T., & Vogt, H. P. (2000). Philippine coral reefs under threat: Lessons learned after 25 years of community-based reef conservation. *Marine Pollution Bulletin, 40*(6), 537–550. https://doi.org/10.1016/S0025-326X(99)00243-X

Wilkinson, C. & Souter, D. (2008). Status of Caribbean coral reefs after bleaching and hurricanes in 2005. Global Coral Reef Monitoring Network, and Reef and Rainforest Research Centre, Townsville, 152 p. (ISSN 1447 6185).

Wynveen, C. J., Connally, W. D., & Kyle, G. T. (2013). Pro-environmental behavior in marine protected areas: The cases of the Great Barrier Reef Marine Park and the Florida keys national marine sanctuary. *Journal of Park and Recreation Administration, 31*(2). https://js.sagamorepub.com/jpra/article/view/3884

Yin, R. K. (2009). *Case study research: Design and methods* (4th ed.). Sage Publications, Inc.

Yusuf, J.-E., Neill, K., St John III, B., Ash, I. K., & Mahar, K. (2016). The sea is rising… but not onto the policy agenda: A multiple streams approach to understanding sea level rise policies. *Environment and Planning C: Government and Policy, 34*, 228–243. https://doi.org/10.1177/0263774X15614457

Zahariadis, N. (2007). The multiple streams framework structure, limitations, prospects. In P. A. Sabatier (Ed.), *Theories of the policy process* (2nd ed., pp. 65–92). Westview Press.

Part 3

SUMMARY AND CONCLUSIONS

Chapter 10

SUMMARY OF CASE STUDIES AND THE CHANGES TO STAKEHOLDER-DRIVEN, PARTICIPATORY MANAGEMENT OF REEFS

By Sabine Bailey and Kelly Dunning

Stakeholder-driven Participatory Management Highlighted through Case Studies

Worldwide, coral reef management is increasingly becoming more participatory and drawing on adaptive management. As a reminder, participatory management refers to a form of management that shares the authority and responsibility of managing natural resources between the government and local communities and stakeholder groups (Kar, 2021). An important aspect is that it incorporates participatory governance which promotes and empowers the people (citizens and non-citizens alike) to participate in the decision-making process (). Often, only entities that significantly affect or are significantly affected by the proposed decision participate. Throughout this book, we've defined these entities as stakeholders (Decker et al., 2012). Our five case studies showcased stakeholder-driven policy-making at the national level through bipartisan legislation in the United States, at the sub-national level through participatory management in Florida, and at the local level through grassroots movements in Florida and the Cayman Islands, with Florida reefs depicted in Figure 10.1.

Our case study on bipartisan legislation in the United States showed that Republicans and Democrats are using coral reefs as symbols in policy-making to collaborate and cooperate on legislation regarding climate-vulnerable ecosystems and the impacts from climate change. For example, coral reefs are symbols of the unique cultural settings of American states across the nation. For Hawaii and Florida, coral reefs represent unique identities and are tied

Figure 10.1 Nurse shark and a school of bluestriped grunts at Pickles Reef off Key Largo, Florida. Photograph by Sabine Bailey.

to their livelihoods and way of life. We also find that there is bipartisan consensus on the need to protect and conserve coral reef ecosystems through community-based management programs funded and supported by the federal government. This means that Congressional leaders are buying into the importance of enabling local communities to play a role in stewarding their coral reefs.

Our case study on the management of Florida's coral reefs examined two organizations: the Florida Keys National Marine Sanctuary (FKNMS) and the Southeast Florida Coral Reef Initiative (SEFCRI). We found that both organizations enacted participatory management styles via stakeholder engagement and involvement. We also found evidence of iteratively learning, or learning as they go, from both organizations, indicating adaptive management strategies. Finally, we identified that the FKNMS engages in a formal and centralized approach to managing the Florida Reef Tract, while SEFCRI engages in an informal and collaborative approach to managing Florida's northern reefs. These cases show how even on the same stretch of reef, management institutions can vary dramatically in the American setting. Thus, it is more important that general principles, such as the use of the best available science and leveraging community participation, inform management rather than a one-size-fits-all management approach.

Our case studies in Miami, Florida and the Cayman Islands examined how local communities organized to contest major port infrastructure projects adjacent to coral habitat and reefs. We also studied how communities stand to lose unique cultural benefits when coral reefs are lost. Across these case studies, major environmental disputes arose. We found that a lack of good governance (i.e., participation, transparency and accountability) by the actors implementing the projects led to the rise of private actors, such as multinational cruise companies and environmental nonprofits, increasing their influence as decision-makers and demanding policy change. Both cases also highlighted the need for greater participation, communication and transparency between all stakeholders involved. The breadth and depth of knowledge that people hold on the cultural services on their reefs demonstrate the level of attachment that many coastal communities have to their local reefs and the need to protect these systems that transcends economic value.

Gaps That Remain: Is Participatory Management Enough?

Divers at Peleliu Coral Gardens Reef in Palau. Photograph by Sabine Bailey.

There is still much to be discovered on how to best implement participatory management in conserving our natural resources. What are the most effective strategies to engage stakeholders and involve them in the decision-making process? How much is feasible and helpful? How can we ensure that legislation that protects coral reefs is passed into law? How do we match scales across national, local and state levels to implement environmental policies to protect our reefs? The answers to these questions will vary from country to country, region to region and even reef to reef. To accommodate global institutional and ecological differences, we argue that participatory management is essential to managing our reefs.

Most prominently, in recent decades and as highlighted in these case studies, we are witnessing a change in favor of including stakeholder participation and adapting to novel ecosystems to restore our reefs. Local communities are coming together to uphold our strong conservation laws by contesting multimillion-dollar projects that may have severe adverse impacts on coral reefs.

However, is this all enough in the face of climate change and global emissions? We argue that it is not. First, we need to acknowledge our role and responsibility toward global carbon emissions and climate change. No matter what local communities do to protect their reefs, they cannot solve the global problem of human-caused climate change. A rapid reduction of greenhouse gas emissions in combination with the management styles that we have described in this book, is what is needed if coral reef ecosystems are to exist in the next 100 years.

In addition, coral reefs occur in the Global North, with its ample resources, and the Global South, where there are more limited economic resources. It is the duty of countries in the Global North to collaborate with countries of the Global South to face the challenges of climate change in a collaborative way. Emissions in the Global North have created the current climate change situation, with impacts borne by the Global South. Shared learning and shared resources are necessary among all countries with coral reefs to enable coral reefs to survive the coming decades.

Bibliography

Ayala-Orozco, B., Rosell, J. A., Merçon, J., Bueno, I., Alatorre-Frenk, G., Langle-Flores, A., & Lobato, A. (2018). Challenges and strategies in place-based multi-stakeholder collaboration for sustainability: Learning from experiences in the global south. *Sustainability (Switzerland)*, *10*(9). https://doi.org/10.3390/su10093217

Ayers, A. L., & Kittinger, J. N. (2014). Emergence of co-management governance for Hawai'i coral reef fisheries. *Global Environmental Change*, *28*(1), 251–262. https://doi.org/10.1016/j.gloenvcha.2014.07.006

Bernauer, T., & Betzold, C. (2012). Civil society in global environmental governance. *Journal of Environment and Development*, *21*(1), 62–66. https://doi.org/10.1177/1070496511435551

Carreira, V., Machado, J. R., & Vasconcelos, L. (2016). Engaging citizen participation—A result of trusting governmental institutions and politicians in the Portuguese democracy. *Social Sciences*, *5*(3), 40. https://doi.org/10.3390/SOCSCI5030040

Chester, M. V, Markolf, S., & Allenby, B. (2019). Infrastructure and the environment in the Anthropocene. *Journal of Industrial Ecology*, *23*(5), 1006–1015. https://doi.org/10.1111/jiec.12848

Coghlan, D., & Brydon-Miller, M. (2014). Participatory governance. In *The SAGE encyclopedia of action research*. https://doi.org/10.4135/9781446294406.N264

Dunlap, R. E., & Mertig, A. G. (1991). The evolution of the U.S. environmental movement from 1970 to 1990: An overview. *Society and Natural Resources*, *4*(3), 209–218. https://doi.org/10.1080/08941929109380755

Dunning, K. H. (2021). Adaptive governance: The proposed port expansion in the Cayman Islands and its impacts to coral reefs. *Marine Policy*, *124*, 104248. https://doi.org/10.1016/j.marpol.2020.104248

Ferse, S. C. A., Glaser, M., Neil, M., & Schwerdtner Máñez, K. (2014). To cope or to sustain? Eroding long-term sustainability in an Indonesian coral reef fishery. *Regional Environmental Change*, *14*(6), 2053–2065. https://doi.org/10.1007/s10113-012-0342-1

Gershtenson, J., Smith, B. W., and Mangun, W. R. (2006). Friends of the earth? Partisanship, party control of congress, and environmental legislation in congress. *Politics & Policy*, *34*(1), 66–92. https://doi.org/10.1111/j.1747-1346.2006.00004.x

Hejny, J. (2018). The Trump administration and environmental policy: Reagan Redux? *Journal of Environmental Studies and Sciences*, 8(2), 197–211. https://doi.org/10.1007/s13412-018-0470-0

Imperial, M. T., & Yandle, T. (2005). Taking institutions seriously: Using the IAD framework to analyze fisheries policy. *Society and Natural Resources*, *18*(6), 493–509. https://doi.org/10.1080/08941920590947922

Jager, N. W., Newig, J., Challies, E., & Kochskämper, E. (2020). Pathways to implementation: Evidence on how participation in environmental governance impacts on environmental outcomes. *Journal of Public Administration Research and Theory*, *30*(3), 383–399. https://doi.org/10.1093/jopart/muz034

Kar, D. (2021). Aspects of participatory management of water bodies. *Community-Based Fisheries Management*, 111–119. https://doi.org/10.1016/B978-0-12-821723-8.00004-X

Koppenol, D. M. (2014). How to settle conflicts during port development. The case of the port of Rotterdam (1973–2008). *Seaports in Transition*, 2071–2079. http://hdl.handle.net/1765/77424

Lawer, E. T. (2019). Examining stakeholder participation and conflicts associated with large scale infrastructure projects: The case of Tema port expansion project, Ghana. *Maritime Policy and Management*, *46*(6), 735–756. https://doi.org/10.1080/03088839.2019.1627013

Lubell, M., Stacey, M., & Hummel, M. A. (2021). Collective action problems and governance barriers to sea-level rise adaptation in San Francisco Bay. *Climatic Change*, *167*(3–4), 1–25. https://doi.org/10.1007/s10584-021-03162-5

Manfredo, M. J., Vaske, J. J., & Sikorowski, L. (2021). Human dimensions of wildlife management. In L. Steg, A. E. van den Berg, & J. I. M. de Groot (Eds.), *Natural resource management: The human dimension* (2nd rev. ed., No. November, pp. 53–72). Johns Hopkins University Press. https://doi.org/10.4324/9780429039706-6

Mease, L. A., Erickson, A., & Hicks, C. (2018). Engagement takes a (fishing) village to manage a resource: Principles and practice of effective stakeholder engagement. *Journal of Environmental Management*, *212*, 248–257. https://doi.org/10.1016/J.JENVMAN.2018.02.015

NOAA. (2021, April 23). Restoring seven iconic reefs: A mission to recover the coral reefs of the Florida keys. NOAA Fisheries. https://www.fisheries.noaa.gov/southeast/habitat-conservation/restoring-seven-iconic-reefs-mission-recover-coral-reefs-florida-keys

Pallis, A. A. (2020). Chapter 3.1 - Port governance and reform | Port economics, management and policy. In *Port economics, management and policy*. https://porteconomicsmanagement.org/?page_id=159

Parvin, P. (2017). Democracy without participation: A new politics for a disengaged era. *Res Publica 2018*, *24*(1), 31–52. https://doi.org/10.1007/S11158-017-9382-1

Prewitt, K., Mackie, C. D., & Habermann, H. (2014). Civic engagement and social cohesion: Measuring dimensions of social capital to inform policy. In *Civic engagement and social cohesion: Measuring dimensions of social capital to inform policy*. https://doi.org/10.17226/18831

Restoring Resilient Reefs Act of 2019, H.R.4160, 116th Cong. (2019). https://www.congress.gov/bill/116th-congress/house-bill/4160

Shah, E., Vos, J., Veldwisch, G. J., Boelens, R., & Duarte-Abadía, B. (2019). Environmental justice movements in globalising networks: A critical discussion on social resistance against large dams. *Journal of Peasant Studies*. https://doi.org/10.1080/03066150.2019.1669566

Smith, Z. A. (2004). *The environmental policy paradox*. Prentice Hall.

Stalley, P., & Yang, D. (2006). An emerging environmental movement in China? *China Quarterly*, *186*(185), 333–356. https://doi.org/10.1017/S030574100600004X

Wiesmeth, H. (2020). Stakeholder engagement for environmental innovations. *Journal of Business Research*, *119*, 310–320. https://doi.org/10.1016/j.jbusres.2018.12.054

Wilbanks, T. J., & Fernandez, S. (2014). Framing climate change implications for infrastructures and urban systems. In *Climate change and infrastructure, urban systems, and vulnerabilities* (pp. 17–40). Island Press/Center for Resource Economics. https://doi.org/10.5822/978-1-61091-556-4_3

Chapter 11

CALL TO ACTION: PARTICIPATION AND MANAGING REEFS UNDER GLOBAL CHANGE

By Kasen Wally and Kelly Dunning

This final chapter will offer three key policy recommendations. The first is a more rapid international timeline for emissions reduction to prevent the extinction of coral that make up reefs due to climate change. This will require bipartisan relationships that mirror those being formed in Congress around coral reef legislative topics. The second is increased cooperation between authorities within federal and state governments to ensure that existing conservation laws and policies are followed, in contrast to the Miami and Cayman Islands cases. The third is that private multinational companies should not be allowed to subvert the conservation laws and policies of sovereign countries, as seen in the case of the Cayman Islands.

In 2022, the United Nations' Intergovernmental Panel on Climate Change (IPCC) released its Sixth Assessment Report, an extensive document examining the ongoing effects of global climate change. Among the key findings of the report lies an ominous warning:

> *The cumulative scientific evidence is unequivocal: Climate change is a threat to human well-being and planetary health. Any further delay in concerted anticipatory global action [...] will miss a brief and rapidly closing window of opportunity to secure a liveable and sustainable future for all* (Pörtner et al., 2022).

The effects of climate change are the most significant threat facing coral reefs, driving their decline, and endangering the communities that depend upon them (International Union for Conservation of Nature, 2021). How can policy-makers rise to the broader challenge of managing coral reefs under the threat of climate change? This question is complicated and, as such, has become the focus of hundreds, if not thousands, of research projects aimed at understanding coral reef management in the Anthropocene. Some studies have found that returning to historical baselines for coral reefs may be impossible and that local management systems are insufficient to rescue them (Bellwood et al., 2019; Hughes et al., 2017). In response, researchers have suggested implementing multilevel, international initiatives such as the Coral Triangle Initiative in Southeast Asia or polycentric governance (or governance systems with many power centers) of the Great Barrier Reef in Australia to promote large-scale ecosystem recovery (Fidelman et al., 2014; Morrison, 2017). Other policy responses include government assistance for livelihoods and building capacity for local communities to adapt to change (Hoegh-Guldberg et al., 2018). Overall, public engagement, flexible institutions that enable managers to adapt to surprises and multi-scale governance with national, state and local authorities working together to respond to change are what is needed to manage coral reef ecosystems.

The underlying problem remains—climate change. As such, we argue for the first of three action items: a more rapid international timeline for emissions reduction to prevent the extinction of coral reefs due to climate change. This recommendation is supported by the findings of the IPCC, which has outlined multiple pathways (i.e., scenarios) that may limit global warming to 1.5°C, a critical threshold that is still possible to meet, but quickly becoming less feasible (Masson-Delmotte et al., 2018). To meet this threshold, these reports emphasize rapid global transformations in infrastructure, including improvements to industrial systems, urban transportation, urban construction and clean energy on a scale never seen before (Masson-Delmotte et al., 2018; Pörtner et al., 2022). It appears that there must be international cooperation to invest in a broad portfolio of mitigation strategies at an unprecedented rate and on a massive scale. Afterall, while we geopolitically divide our oceans and natural resources, there is only *one* global ocean, which we all must share and cooperate to protect. The same can be said for the global climate and terrestrial environment.

Additionally, the IPCC states that it is imperative that these strategies to improve infrastructure and emissions timelines must also target the restoration of natural ecosystems, safeguard biodiversity and ensure ecological resilience wherever possible (Pörtner et al., 2022). The ecosystem services provided by these natural resources, like the benefits of coral reefs, often underpin carbon sequestration and cycling (Pörtner et al., 2022). For instance, while coral reefs are not a net carbon sink (i.e., a net absorber of CO_2), they directly support seagrass beds, mangrove forests and wetlands, which play a significant role in carbon sequestration (UNFCCC, 2021).

Our second policy recommendation is to increase cooperation between authorities within federal and state governments to ensure that existing conservation laws and policies are followed. Contemporary research has clarified that adaptive strategies are necessary to save coral reefs and the people who depend upon them (Cinner et al., 2016; Comte & Pendleton, 2018; Morrison et al., 2020). Within these adaptive management strategies for coral reefs, cooperation is absolutely essential at the international, federal, state and local levels. However, one of the most significant issues is that of scale; coral reef ecosystems are often large, and their decline is rapidly increasing, making conservation difficult to implement and time-sensitive. In areas like the Coral Triangle, where over 100 million people span six countries and depend on coral reefs, multi-scale cooperation is critical, especially involving local communities (Dunning, 2022). Thus, management must find common ground between governmental and community-based interventions. To achieve this, we know that management must incorporate local knowledge into regional governance while considering the reciprocal linkages between human activities, ecosystem services and human well-being (Dunning, 2022; Kittinger, 2012; Kittinger et al., 2016). Indeed, the effects of these management approaches are being seen throughout the world, from the Coral Triangle to the Florida Keys and to the Red Sea (Dunning, 2022; Fine et al., 2019; Gregg et al., 2013). However, because this book has focused heavily on coral reef conservation within the federalist system of the United States (i.e., the case of coral reefs in Congress, the case of the Florida Reef Tract, and the case of the Port of Miami expansion project), we urge that in addition to cooperative management solutions elsewhere, there must also be a focus on cooperation between federal and state governments.

As we saw in our comparative case study of the Florida Keys National Marine Sanctuary (FKNMS) and the Southeast Florida Coral Reef Initiative (SEFCRI), understanding the dynamics of federal and state governments within regional ecosystems (which include different types of institutions, actors, and management styles) can help to improve conservation. For instance, we saw that SEFCRI emerged to fill a functional role, generating a complete understanding of the northern reaches of the Florida Reef Tract, which until 2003 lacked any cohesive management. SEFCRI's existence has enabled management to begin transitioning toward formal and regulatory governance (like that of the FKNMS), such as a new policy priority being placed on environmental regulations for the Kristin Jacobs Coral Reef Ecosystem Conservation Area (i.e., The SEFCRI Region). This transition away from novel data collection, toward formal and regulatory management, may ultimately homogenize governance, making cooperation between the two entities easier at the regional scale. Likewise, we saw that the rigid, hierarchical structure of the FKNMS could perhaps benefit from some elements of SEFCRI's broad actor network, drawing on increasingly diverse backgrounds to generate novel solutions to anthropogenic threats. These types of formal-informal linkages could be one way to encourage cooperation within the federal and state government. However, more importantly it addresses cooperative management within a complex regional ecosystem, a scenario that involves multiple jurisdictions, and is often difficult for managers to navigate.

Regional governance, especially in the context of large coral reef ecosystems, like the Florida Reef Tract, The Great Barrier Reef, or the Mesoamerican Reef, draws interesting parallels to the governance of other water resources, like those found in terrestrial watersheds. The reason being that the ecosystems in question are transboundary and involve complex landscapes of administrative boundaries and various types of governance (Pahl-Wostl, 2020). Thus, multi-scale governance is required to coordinate management effectively (e.g., local municipalities must coordinate with multiple state agencies). Achieving this type of governance requires high levels of coordination (i.e., cooperation) among federal, state and local actors. Additionally, there must be a balance between top-down and bottom-up authority, as witnessed via our comparative case study of the FKNMS and SEFCRI (Pahl-Wostl, 2020). This system of distributed management with many power centers is called polycentric governance. If increased cooperation is necessary to achieve regional management, we suggest that further research and management strategies start there. By doing so, future management can build on polycentric governance frameworks, like that of Carlisle and Gruby (2019), who emphasize the applicability of polycentric governance in transboundary natural resource management, and suggest that without cooperation, isolated and centralized governance systems are prone to failure. Thus, decentralized multi-scale governance offers a way to transcend conventional management and form a united front to fight climate change and help to save coral reefs.

Despite the best management strategies, external factors, such as special interests in the private sector, can undermine conservation in pursuit of economic gains or other favorable outcomes for the business. For instance, as we saw in the case of the Cayman Islands port expansion project, the ability to dock cruise ships rather than tender passengers to and fro has resulted in a significant threat to a portion of renowned and beloved coral reef habitat. Given these complications, which can be highly contentious among opposing actors (e.g., the cruise industry and Caymanian people) and potentially disastrous for coral reefs, let us turn to our third policy recommendation: private multinational companies should not be allowed to subvert the conservation laws and policies of sovereign nations.

It is no secret that multinational companies exercise considerable power over the biosphere and its natural resources, including coral reefs. Many of the resources that we use every day are held in the hands of a few multinational corporations; for example, the majority of the world's most valuable fisheries (both economically and environmentally) are dominated by just 13 companies (Folke et al., 2019). Additionally, these inequalities are not only abundant in fisheries but also in agriculture, forestry, fertilizers, minerals and fossil fuels (Folke et al., 2019). So, what sorts of responsibilities do these companies have to participate in global conservation efforts in the twenty-first century?

The concept of corporate responsibility, defined as "the expectation that private companies should no longer base their actions on the needs of their shareholders alone, but be more broadly responsive to the needs of the society in which they operate," has been explored in the environmental context (Morgera, 2020, p. 16). Unfortunately, multinational corporations' roles in identification and protection of site-specific ecosystems remain weak (like the threatened reefs of Cayman Islands) (Morgera, 2020, pp. 224–226). While there has been an emphasis placed on developing international oversight to monitor the behavior of multinational corporations, there remains a literature gap on corporate environmental responsibility and coral reefs. We suggest that while it is well-known that multilevel cooperative management among decision-makers and stakeholders is beneficial (and often necessary) for coral reef governance, it remains to be seen how corporate environmental responsibility and accountability fit into this frame. What happens in situations where some political leaders enable private companies to take actions that violate domestic conservation laws? Is the type of situation that we have detailed in the Cayman Islands happening in other places in other ecological settings as well?

When countries implement strict environmental regulations, corporations are likely to find ways to circumvent those requirements. In the case of CO2 emissions, Ben-David et al. (2018) found that corporations responsible for polluting emissions tended to relocate their activities to countries where regulations were less stringent or more challenging to enforce. With these types of nefarious activities taking place, it becomes even more important that sovereign nations strengthen their environmental regulations and that international oversight continues to be generated for multinational companies. However, this may be easier said than done. Even in places like the United States, where sweeping environmental laws (e.g., the Clean Air Act) have helped regulate unsustainable business practices, rollbacks during unfavorable presidential administrations have reduced corporate environmental accountability (Hayes, 2020a). For instance, the Trump administration reduced decades-old protections under the Clean Water Act, potentially affecting up to 50 percent of U.S. wetlands and 20 percent of streams (Hayes, 2020b).

While there may not be an easy way to fix these issues, some of which remain pernicious (even under well-established institutions), there is a way forward. When corporations recognize biodiversity loss and environmental decline as affecting their bottom line, incentives begin to appear (Dempsey, 2013). By focusing socio-political movements on environmental concerns, pressure can be applied via public demand for new regulations and financial incentives that target big businesses (Addison & Bull, 2018; Redford et al., 2015). Ultimately, by holding corporations accountable, we can achieve better outcomes for coral reefs and better outcomes for the planet.

In conclusion, multi-scale cooperation between communities, state and federal governments, private businesses and international partners can underpin policy recommendations as we have proposed here. For instance, these sorts of collaborations can support a more rapid international timeline for emissions reduction and, in doing so, help to prevent the decline of coral reefs due to climate change. Additionally, these sorts of multi-scale collaborations foster a unified effort to protect Earth's natural resources in a way that carefully considers the human dimensions of the problem and recognizes coral reefs as a critical resource for the people. Through dedicated stewardship guided by sound science and robust policy, there remains hope for the people's reefs.

References

Addison, P. F., & Bull, J. W. (2018). Conservation accord: Corporate incentives. *Science, 360*(6394), 1195–1196.

Bellwood, D. R., Pratchett, M. S., Morrison, T. H., Gurney, G. G., Hughes, T. P., Álvarez-Romero, J. G., Day, J. C., Grantham, R., Grech, A., Hoey, A. S., Jones, G. P., Pandolfi, J. M., Tebbett, S., Techera, E., Weeks, R., & Cumming, G. S. (2019). Coral reef conservation in the Anthropocene: Confronting spatial mismatches and prioritizing functions. *Biological conservation, 236*, 604–615.

Ben-David, I., Kleimeier, S., & Viehs, M. (2018). *Exporting pollution: Where do multinational firms emit CO2? (No. w25063)*. National Bureau of Economic Research.

Carlisle, K., & Gruby, R. L. (2019). Polycentric systems of governance: A theoretical model for the commons. *Policy Studies Journal, 47*(4), 927–952.

Cinner, J. E., Huchery, C., MacNeil, M. A., Graham, N. A., McClanahan, T. R., Maina, J., Maire, E., Kittinger, J. N., Hicks, C. C., Mora, C., & Allison, E. H. (2016). Bright spots among the world's coral reefs. *Nature, 535*(7612), 416–419.

Comte, A., & Pendleton, L. H. (2018). Management strategies for coral reefs and people under global environmental change: 25 years of scientific research. *Journal of Environmental Management, 209*, 462–474. https://doi.org/10.1016/j.jenvman.2017.12.051

Dempsey, J. (2013). Biodiversity loss as material risk: Tracking the changing meanings and materialities of biodiversity conservation. *Geoforum, 45*, 41–51.

Dunning, K. (2022). Biodiversity conservation policy in megadiverse countries: Comparing policy systems for 2020 targets to inform management in the coming decades. *Journal of Environmental Management, 302*, 113815.

Fidelman, P., Evans, L. S., Foale, S., Weible, C., Von Heland, F., & Elgin, D. (2014). Coalition cohesion for regional marine governance: A stakeholder analysis of the coral triangle initiative. *Ocean & Coastal Management, 95*, 117–128.

Fine, M., Cinar, M., Voolstra, C. R., Safa, A., Rinkevich, B., Laffoley, D., Hilmi, N., & Allemand, D. (2019). Coral reefs of the Red Sea—Challenges and potential solutions. *Regional Studies in Marine Science, 25*, 100498.

Folke, C., Österblom, H., Jouffray, J. B., Lambin, E. F., Adger, W. N., Scheffer, M., Crona, B. I., Nyström, M., Levin, S. A., Carpenter, S. R., & Anderies, J. M. (2019). Transnational corporations and the challenge of biosphere stewardship. *Nature Ecology & Evolution, 3*(10), 1396–1403.

Gregg, K. (2013). Management considerations for the Southeast Florida coral reef ecosystem. https://floridadep.gov/sites/default/files/FDOU_Management_Considerations_0.pdf

Hayes, D. (2020a, February 20). Reducing corporate environmental accountability, whether industry wants it or not. New York University School of Law. Retrieved April 12, 2022, from. https://www.law.nyu.edu/centers/state-impact/blog/reducing-corporate-environmental-accountability

Hayes, D. (2020b, February 6). Ideology Trumps science in clean water rollback. New York University School of Law. Retrieved April 12, 2022, from https://www.law.nyu.edu/centers/state-impact/blog/clean-water-rollback

Hoegh-Guldberg, O., Kennedy, E. V., Beyer, H. L., McClennen, C., & Possingham, H. P. (2018). Securing a long-term future for coral reefs. *Trends in Ecology & Evolution, 33*(12), 936–944.

Hughes, T. P., Barnes, M. L., Bellwood, D. R., Cinner, J. E., Cumming, G. S., Jackson, J. B., Kleypas, J., van de Leemput, I. A., Lough, J. M., Morrison, T. H., Palumbi, S. R., van Nes, E. H., & Scheffer, M. (2017). Coral reefs in the Anthropocene. *Nature, 546*(7656), 82–90.

International Union for Conservation of Nature (IUCN). (2021, March 2). *Coral reefs and climate change*. IUCN. Retrieved March 18, 2022, from https://www.iucn.org/resources/issues-briefs/coral-reefs-and-climate-change

Kittinger, J. N., Bambico, T. M., Minton, D., Miller, A., Mejia, M., Kalei, N., Wong, B., & Glazier, E. W. (2016). Restoring ecosystems, restoring community: Socioeconomic and cultural dimensions of a community-based coral reef restoration project. *Regional Environmental Change, 16*(2), 301–313.

Kittinger, J. N., Finkbeiner, E. M., Glazier, E. W., & Crowder, L. B. (2012). Human dimensions of coral reef social-ecological systems. *Ecology and Society, 17*(4).

Masson-Delmotte, V., Zhai, P., Pörtner, H.-O., Roberts, D., Skea, J., Shukla, P. R., Pirani, A., Moufouma-Okia, W., Péan, C., Pidcock, R., Connors, S., Matthews, J. B. R., Chen, Y., Zhou, X., Gomis, M. I., Lonnoy, E., Maycock, T., Tignor, M., & Waterfield, T. (Eds.). (2018). IPCC 2018: Summary for policymakers. In *Global warming of 1.5°C. An IPCC special report on the impacts of global warming of 1.5°C above pre-industrial levels and related global greenhouse gas emission pathways, in the context of strengthening the global response to the threat of climate change, sustainable development, and efforts to eradicate poverty*. IPCC. World Meteorological Organization, 32 pp. https://www.ipcc.ch/sr15/chapter/spm/

Morgera, E. (2020). *Corporate accountability in international environmental law* (No. 2nd). Oxford University Press.

Morrison, T. H. (2017). Evolving polycentric governance of the Great Barrier Reef. *Proceedings of the National Academy of Sciences, 114*(15), E3013–E3021.

Morrison, T. H., Adger, N., Barnett, J., Brown, K., Possingham, H., & Hughes, T. (2020). Advancing coral reef governance into the Anthropocene. *One Earth, 2*(1), 64–74.

Pahl-Wostl, C., Knieper, C., Lukat, E., Meergans, F., Schoderer, M., Schütze, N., Schweigatz, D., Dombrowsky, I., Lenschow, A., Stein, U., Thiel, A., Tröltzsch, J., & Vidaurre, R. (2020). Enhancing the capacity of water governance to deal with complex management challenges: A framework of analysis. *Environmental Science & Policy, 107*, 23–35.

Pester, P. (2021, December 12). When did scientists first warn humanity about climate change? *Livescience.Com*. Retrieved March 18, 2022, from https://www.livescience.com/humans-first-warned-about-climate-change

Pörtner, H.-O., Roberts, D. C., Poloczanska, E. S., Mintenbeck, K., Tignor, M., Alegría, A., Craig, M., Langsdorf, S., Löschke, S., Möller, V., & Okem, A. (Eds.). (2022). IPCC, 2022: Summary for policymakers. In *Climate change 2022: Impacts, adaptation, and vulnerability. Contribution of working group II to the sixth assessment report of the intergovernmental panel on climate change*. Cambridge University Press. In Press. https://www.ipcc.ch/report/ar6/wg2/downloads/report/IPCC_AR6_WGII_SummaryForPolicymakers.pdf

Redford, K. H., Huntley, B. J., Roe, D., Hammond, T., Zimsky, M., Lovejoy, T. E., da Fonseca, G. A., Rodriguez, C. M., & Cowling, R. M. (2015). Mainstreaming biodiversity: Conservation for the twenty-first century. *Frontiers in Ecology and Evolution*, 137.

United Nations Framework Convention on Climate Change (UNFCCC). (2021, October 11). Oceans and coastal zones | Resilience | Coral reefs. United Nations: Climate Change. Retrieved April 5, 2022, from https://unfccc.int/blog/coral-reefs

APPENDIX CHAPTER 5

Methods

Data included meeting minutes (i.e., transcripts recorded during advisory council/committee meetings), news documents (including newspaper articles, press releases, newspaper blogs and news transcripts), and remote interviews with representatives and participating stakeholders from the Southeast Florida Coral Reef Initiative and the Florida Keys National Marine Sanctuary. Using the appearance of mass coral bleaching events along the Florida Reef Tract as a temporal threshold, denoting a shift from historical to contemporary ecosystem functionality and management needs, I compiled a total of 1,122 individual data points (n = 1,122) from both organizations, FKNMS (n = 541) and SEFCRI (n = 581), throughout the summer of 2021.

News document data were collected from the Nexis Uni database, which initially revealed over 10,000 results when searching "Florida Keys National Marine Sanctuary" or "Southeast Florida Coral Reef Initiative." However, many of these results were irrelevant within the scope of this study; therefore, modifiers were added to narrow the search results. First, for the Florida Keys National Marine Sanctuary, 179 documents were isolated using the "News" filter and the Boolean search "Florida Keys National Marine Sanctuary" AND "management." These results were then filtered using the category "Newspapers" and the date range "1996 - present." The reason I focused on newspaper articles was to make the amount of FKNMS news data more manageable. Additionally, 1996 was selected as a starting point because the first mass bleaching event to occur in Florida after the creation of the FKNMS was in 1997. Of the 179 results, 79 were deemed relevant to this study, and 100 were excluded. The results were manually sorted to determine their relevance to this research. Any articles that were outside the scope of this study were therefore excluded. For example, one article focused on new funding for habitat restoration in the Everglades National Park and mentioned both the Florida Keys National Marine Sanctuary and "management" in that context. However, this had no direct relevance to the management of the Florida Reef Tract (Wadlow, 2010).

Second, for the Southeast Florida Coral Reef Initiative, 99 results were isolated using the "News" filter and the Boolean search "Southeast Florida Coral Reef Initiative" OR "SEFCRI" OR "Ecosystem Conservation Area." Because SEFCRI is a newer organization (established in 2003), there was no need to limit the date range. However, the more recent establishment of this organization also meant that there were fewer newspaper articles available. Therefore, I also included more comprehensive sources of news such as the newswires, press releases, newspaper blogs and news transcripts previously mentioned. For this reason, the decision was made to manually sort all 99 initial results for relevance to this project rather than narrow the results any further using modifiers or filters. Of the initial 99 results, 20 were deemed relevant, and 79 were excluded. These decisions were based on similar criteria used for the FKNMS results (i.e., relevance to the management of the Florida Reef Tract). For example, one recent newspaper article announced the renaming of SEFCRI's "Coral Reef Ecosystem Conservation Area" to the "Kristin Jacobs Coral Reef Ecosystem Conservation Area." However, this information was deemed irrelevant to SEFCRI's management of the Florida Reef Tract (Nicol, 2021).

Meeting Minutes data from each organization's advisory meetings were collected from two sources. First, the meeting minutes for the Florida Keys National Marine Sanctuary's Sanctuary Advisory Council were accessed via the archives on their website (Florida Keys National Marine Sanctuary, n.d.-a). Second, the meeting minutes for the Southeast Florida Coral Reef Initiative's Technical Advisory Committee were accessed via the archives on the Florida Department of Environmental Protection's website (Florida Department of Environmental Protection, 2022). I chose meeting minutes from the Sanctuary Advisory Council and Technical Advisory Committee because these groups provided a mix of representatives from each organization and stakeholders from different communities associated with the Florida Reef Tract. While the Sanctuary Advisory Council and Technical Advisory Committee lack direct rule-making authority, they each play critical roles within the decision-making processes of each organization and are therefore highly representative (Florida Department of Environmental Protection, n.d.-d; Florida Keys National Marine Sanctuary, n.d.-b). To maximize comparability between the two organizations, I chose only to select meeting minutes available from each organization at similar points in time (Spring/Summer, 2015–2020). This method resulted in a total of 12 meeting transcripts, one transcript per organization per year.

Remote interview data were collected from transcripts of semi-structured interviews conducted throughout the summer field season (May 5–August 1) of 2021. Participants were selected based on their status as important stakeholders, decision-makers and representatives from the SEFCRI Team Charter

(9 participants) and the FKNMS Advisory Council (9 participants). The semi-structured interview methodology used in this research project was approved by the Office of Human Research at Auburn University (Protocol # 21-088 EP 2103). Written consent was obtained from each participant before each remote interview. The consent form included (1) a brief introduction to the study; (2) information related to how the data would be used; (3) an explanation of any risk involved to the participants; (4) all steps taken to ensure risk was minimized; and (5) information about the participant's right to withdraw from the study at any point in time voluntarily. To help ensure honest and representative answers, participants' personal information remained completely confidential at all times. Interviews were conducted virtually using Zoom™ software. At the beginning of each interview, participants were asked permission to record audio and video logs which the Zoom software automatically transcribed. These transcripts were then double-checked by hand to ensure accuracy. If the participant agreed, Zoom recordings and transcripts were saved to a secure Auburn University cloud account. If the participant preferred not to be recorded, detailed notes were taken by hand for the duration of the interview and stored securely.

After all data sources had been compiled, I began a process of deductive qualitative coding using the five concepts for managing novel ecosystems identified by Clement and Standish (2018) and loop-learning as defined in Pahl-Wostl's Framework for Adaptive Governance (2009). Results were reviewed and coded based on their content related to (1) Issue Framing, with a focus on reframing the problem and adopting novel solutions; (2) Culture and Norms with an emphasis on stakeholder inclusion and cultural values; (3) Power and Authority, including information on clearly defined roles/responsibilities and perceived legitimacy; (4) Administrative Competence, assessing whether practitioners are effectively doing their jobs; and (5) Buffering, which assessed preparedness related to sources of organizational and ecological uncertainty. Additionally, each data point was coded for the presence or absence of single, double and triple-loop learning to help gauge institutional progress toward effective adaptive governance. Finally, the content of each data point was assessed as either negative or positive, denoting the presence or absence of criticism about the organization.

The document analysis process for this study initially began with each document representing one data point which included multiple perspectives from different stakeholders. For example, some news documents, such as newspaper articles, included four or five competing perspectives related to the Administrative Competence of either organization about various concerns. Additionally, these perspectives were often framed in a positive or negative light. Because of these factors, the aggregate data lacked some specificity, which threatened to attenuate my results. To remedy this problem, each document was treated as a single source for multiple data points, each representing different perspectives of stakeholders and decision-makers. Afterward, this information was analyzed and leveraged to compare the two institutions and their approaches to managing the Florida Reef Tract. Below I outline my coding methodology in greater detail.

For this research, I utilized a deductive content analysis approach that used elements of concept coding as described by Saldaña (2016). Deductive content analysis is an analytical method that can test existing theoretical frameworks, categories or concepts in a new context and with new data (Kyngäs & Kaakinen, 2020). In other words, deductive content analysis applies prior theoretical knowledge as the starting point, in this case, the five concepts for managing novel ecosystems, and allows those concepts to act as a lens through which to examine new cases. Additionally, I incorporated concept coding as a way to allow for some flexibility and generality. Concept coding also allows specific statements to be grouped into general categories (i.e., the five important concepts for managing novel ecosystems outlined in the table below) via their related elements (Saldaña, 2016). In this case, I refer to the individual elements of each concept as subcomponents. For example, within Administrative Competence, many different kinds of statements can be categorized based on their content. Two such subcomponents were Feasibility and Technical Know-How. These subcomponents were originally derived from the definition of Administrative Competence provided by Clement and Standish (2018). However, as needed, additional subcomponents were created based on their relevance to themes in the data. For example, Perceived Legitimacy was added as a subcomponent within Power and Authority based on my empirical observations that there existed public contention around the establishment of regulatory bodies like the FKNMS.

Before any data sources were coded, several pieces of information were recorded: the source URL, the date of the document, the date of entry, the specific quote, the speaker's name, the speaker's organizational affiliation, and their role (if any). Then each quote was assessed and coded based on its relevance to each of the five concepts for adaptive governance and its applicability to single, double or triple-loop learning. At the same time, the phrasing and content of each quote were assessed as either "positive" or "negative," and the results were recorded. This was done to identify where the highest levels of criticism exist for each organization, highlighting areas that could be considered for future improvement via adaptive governance and management practices. Lastly, a justification was added for each code, describing why subcomponents were identified. This was done in order to provide additional context and specificity. This process was repeated for every entry in my datasheet. More detailed descriptions and examples are given in my findings section.

For my quantitative analysis, I used two-tailed difference of proportions t-tests to identify significant differences in the proportions of each concept as well as differences in the proportions of single, double, and triple-loop learning. While my purposeful sampling was not completely random or exhaustive, the diversity in my data sources, including news media, advisory meetings and interviews, ensured a large and representative sample of the stakeholders and practitioners involved with both the FKNMS (n = 541) and SEFCRI (n = 581). Once the significant differences had been calculated for each concept and each level of learning, I was able to explore each theme in-depth and couple my qualitative and quantitative data to accurately describe the differences between each organization.

Data quality assurance and quality control (sometimes referred to as QA/QC) were conducted throughout the data collection phase of this project. QA/QC includes data validation, cleaning and consistent checking, which ensures data integrity (Bowser et al., 2020). To do this, I employed a process of inter-coder reliability, where an additional independent coder double-checks the work of the initial coder to identify any irregularities and safeguard consistency within the dataset (Lavrakas, 2008). In addition, I used triangulation, where different sources of data (i.e., news, meeting minutes, and interviews) were combined with quantitative and qualitative analysis (mixed methods) to improve the validity of my research and help control sources of potential bias (Creswell & Miller, 2000).

Codebook for Qualitative Codes

Concept	*Definition*	*Criteria**
		*** (1) indicates the presence of the concept; (0) indicates the absence of the concept; (+) indicates a positive mention of the concept; (−) indicates a criticism of the concept**
Administrative Competence	Information about whether management is effectively conserving the ecosystem	(1) The presence of any components related to administrative competence (+) Positive examples include managers incorporating scientific data into their decision-making processes; personnel being adequately trained and knowledgeable; the presence of iterative learning and cooperation; assessments conducted for project feasibility. (−) Criticisms include decisions being made without consideration of scientific information; personnel inadequately prepared or absent from the decision-making process; past mistakes and new information not being used to improve management; projects progressing without cost/feasibility assessments (0) The absence of this concept

(Continued)

(Continued)

Codebook for Qualitative Codes

Concept	*Definition*	*Criteria**
Buffering	Information about whether management is adequately prepared for the future	(1) The presence of any components related to buffering (+) Positive examples include managers responding adequately to environmental changes; managers making decisions at the right times despite conditions of significant uncertainty; managers collecting enough data to inform decision-making; managers considering ways to improve resilience and manage uncertainty (−) Criticisms include insufficient data collection; poor planning; managers not being prepared to implement new decisions when the time comes; managers not considering ways to reduce uncertainty and increase resilience (0) The absence of this concept
Culture and Norms	Information about whether management includes stakeholders, engages the public, and encourages a shared understanding	(1) The presence of any components related to culture and norms (+) Positive examples include broad opportunities for inclusive public discourse and communication; overcoming enduring preferences for conventional management actions (e.g., protected areas, restoration to baselines, protecting specific species); shared objectives among stakeholders (the public, managers, scientists, fishers, developers); reaching common ground between actors (−) Criticisms include failures to disseminate information to the public or decision-makers; failures to foster an inclusive environment which promotes shared ideas and beliefs; a lack of public outreach; programs not meeting their objectives; instances of public confusion. (0) The absence of this criteria
Issue Framing	Information about whether management adequately considers anthropogenic impacts and novel solutions	(1) The presence of any components related to issue framing (+) Positive examples include new ways of framing the problem of biodiversity and ecosystem decline; considering new species interactions, compositions or ecosystem functionality; re-prioritizing objectives; redefining what conservation means in the Anthropocene; describing different types of novel solutions (−) Criticisms include agencies and managers not addressing anthropogenic threats; failures to incorporate novel conditions into management discussion and decision-making (0) The absence of this concept

(*Continued*)

(Continued)

Codebook for Qualitative Codes

Concept	*Definition*	*Criteria**
Power and Authority	Information about whether management is well organized, capable, respected and willing to consider change	(1) The presence of any components related to power and authority (+) Positive examples include the reallocation of rights and responsibilities; new procedures being implemented; roles and responsibilities clearly defined and accepted; feasible opportunities have been identified where redistribution of power can occur; regulations implemented and enforced (−) Criticisms include roles of different organizations are poorly defined; regulations were not enforced; ineffective changes to management plans or institutional structure; instances of mistrust between stakeholders and managers (0) The absence of this concept

APPENDIX CHAPTER 6

Methods

Case selection

The 2015 Port of Miami dredging project in Florida, United States, was the selected case study for this research. An environmental dispute emerged due to the organization of local stakeholders who were concerned over the uncertainty over ecological impacts of these projects on reefs. The project took place in Miami, which is a popular destination in the tropics, whose economy is strongly tied to trade and the tourism industry. Furthermore, the port project takes place in a wealthy location based on gross domestic products and has institutions and resources that allow for coalitions to emerge.[1] We apply the Narrative Policy Framework as the theoretical lens to assess how narrative strategies were used to influence policy outcomes regarding these port projects.

Research design

The case study is the research design for this study, and it is appropriate because it allows the researcher an in-depth investigation to answer why and how decisions were made, examines contemporary events over which the researcher has no control over and enables multiple sources of data to converge toward an explanation (Yin, 2018). The project specifically uses a case-study design and mixed methods to provide more rigorous and robust findings (Creswell & Clark, 2019; Johnson et al., 2007; Yin, 2018). Mixed methods research is when a researcher combines qualitative and quantitative elements of research methods and integrates multiple forms of data

1 2019 USA gross domestic product (GDP) per capita is 65,297.52 USD (WorldBank); 2017 Miami-Fort Lauderdale GDP per capita is 48,140 USD (FRED Economic Data); 2018 Cayman Islands GDP per capita is 85,975 USD (WorldBank); the average GDP per capita as of 2019 is 11,433 USD (WorldBank).

Table A.6.1 Policies, committees and corresponding agencies that protect the coral reefs in Florida.

Policy/Plan/Law	*Agency in Charge*	*Purpose*
Endangered Species Act (1973)	National Marine Fisheries Services	Prohibits destruction of critical habitat and unauthorized harvest of all listed species. Section 7 requires consultation with NMFS for actions concerning staghorn coral (*Acropora cervicornis)* and elkhorn coral (*Acropora palmata).*
Coral Reef Conservation Act (2000)	US Coral Reef Task Force/ National Oceanic and Atmospheric Administration	To preserve and protect coral reef ecosystems, promote effective management and sustainable use, research and be informed by science, and assist programs that involve coral reefs and affected local communities.
Marine Protection, Research, and Sanctuaries Act (1972)	US Army Corps of Engineers/ Environmental Protection Agency	Prohibits the dumping of materials in the ocean that would excessively threaten human health or the marine environment. Under Section 103, USACE issues permit to authorize disposal and dumping of materials in the ocean, upon review by the Environmental Protection Agencies.
Florida Keys National Marine Sanctuary and Protection Act (1990)	Environmental Protection Agency/ National Oceanic and Atmospheric Administration/ Florida Department of Environmental Protection/ Florida Fish and Wildlife Conservation Commission	To protect marine resources, promote education about the Florida Keys to the public, and manage anthropogenic activities in accordance with the Act. The Florida Keys National Marine Sanctuary has strict no-take policies regarding removing, harvesting, taking, injuring coral or live rock and protected wildlife including fish. Discharging materials, operating vessels that will damage marine life and dredging is also strictly prohibited.
Coral Reef Conservation Program (2000)	National Oceanic and Atmospheric Administration	To protect, conserve and restore coral reef resources by maintaining healthy ecosystem function.
Southeast Florida Coral Reef Initiative (2000)	US Coral Reef Task Force/ Florida Department of Environmental/ Florida Fish and Wildlife Conservation Commission Protection	To develop a cooperative strategy with all interested parties to preserve and protect southeast Florida's coral reefs, with an emphasis on the balance between resource use and protection.

(Creswell & Clark, 2019). We collected document and interview data from local stakeholders involved in the port project to identify the narrative strategies that were employed.

Data collection

For the Port of Miami case study, data were collected from four sources: the Nexis Uni database, the US Army Corps of Engineers (USACE), Jacksonville District news archive, the Miami WaterKeeper press and blog archive, and the World Access News database. A total of $n = 103$ documents were collected, which consisted of news releases, articles and position statements. Using the search terms "Miami" and "Deep Dredge Project" between 2010 and 2020 in Nexis Uni resulted in $n = 65$ documents. This date range was chosen because political leaders announced the start of the deep dredge project in 2010, and even though the project was completed in 2015, coral mortality reports and papers were published from 2016 to 2019. We chose 2020 as the cut-off because articles concerning the USACE and port construction were primarily focused on the Port Everglades project, which is the next big port dredging project scheduled in South Florida. We excluded 35 documents from these results because they were either duplicates, focused on the Port Everglades project or merely mentioned the deep dredge project in passing. The included documents were primarily news releases of position statements from various political leaders. The USACE Jacksonville District news archive was chosen because they were the supervisors of the project and thus integral to this research. The news archive provided $n = 19$ articles that were tagged with "Miami" or "Dredging" or "Harbor" ranging from 2012 to 2019, of which $n = 12$ were included for data analysis. Those excluded either were not relevant to the Port of Miami project or focused on the Port Everglades project. The Miami WaterKeeper press and blog archive was chosen because they were one of the organizations that spearheaded the narrative that contested the dredging impacts on reefs. The blog archive collected articles from various news sources, such as the *Miami New Times*, *Miami Herald*, the *New York Times*, dating from 2014 to 2019. The platform had $n = 47$ articles regarding the Port of Miami dredging project, $n = 36$ of which were included for data analysis. The remaining 11 were duplicates of the same press release published across different news outlets. To ensure we had reached information saturation, using the same search terms as the Nexis Uni database, we searched the World Access News database for more articles. There were $n = 83$ results, of which I included $n = 25$ articles as the rest were irrelevant or had already been accounted for.

To ensure narratives from both cases were authentic representations, we restricted analysis to quotations, which we counted as individual data points. Data points (total $n = 211$) from documents were collected until we reached information saturation (Saunders et al., 2018). To supplement and triangulate my document data points, we conducted $n = 10$ semi-structured interviews for the Miami case study, see Table 3.2 for more details (Nightingale, 2020; Price, 2019). To find potential respondents, we used a snowball sampling technique where we asked interviewees to recommend other stakeholders for us to interview (Sadler et al., 2010; Taherdoost, 2018). Interviews were conducted between May 2021 and September 2021. At the beginning of each interview, respondents were asked for permission to record the interview via Zoom. If permission was given, the interview was recorded and transcribed from the recording at a later time. If permission was not given, their responses were transcribed during the interview and edited at a later time, only to improve coherency of the sentences. Respondents were asked nine questions about the costs and benefits of the project, its impact on coral reefs and the use of science regarding the port projects. Interviews ranged from 19 to 229 minutes (interview questions are below). Due to the anonymity and given verbal consent by respondents in these interviews, ethical status was maintained.[2]

Interview Manual

1. What was your role during the Port of Miami Dredging Project that took place from 2013 to 2015?
2. What were/are the benefits of the dredging project?
3. Who is or will be benefiting from them?
4. What were/are the costs of the dredging project?
5. Who is or will be paying for them?
6. In Florida, there are established rules for protecting coral reefs. Did the port dredging project challenge this in any way?
7. How did your organization ensure that science was used to inform the project?
8. How did your organization use social or regular media to portray the project?
9. Did communicating about this project involve significant legal or financial efforts from your organization?

2 Protocol # 21-103 EX 2102.

Table A.6.2 Respondent categories.

Respondent Type	*Examples of Respondents*	*Number of Respondents for the Miami Case*
Non-governmental organizations	Local citizen groups, environmental conservation organizations, local advocacy groups	3
Private sector	Contracting companies, contracted scientists and universities	6
Federal actors	Managers, support staff	1

Table A.6.3 Port of Miami Deep Dredge Project data collection summary.

	Documents	*Data Points*	*Pro-Project*	*Anti-Project*
Nexis Uni	30	46	42	4
USACE News Archive	12	18	16	2
Miami WaterKeeper Blog Archive	36	87	20	65
World Access News	25	50	32	18
Interviews	10	10	2	8
Sum	113	211	114	97

Data coding

Data points and interview transcripts were coded using a codebook based on the Narrative Policy Framework (Shanahan et al., 2016, 2018). We focused primarily on the presence and absence of specific narrative elements and coded 1 for presence and 0 for absence. If the element was present, I used a subcode of the element to keep track of its frequency to aid in quantitative and qualitative analysis later on (Saldaña, 2016). For example, if a hero character was present in a data point, I coded this as 1 and had an accompanying subcode indicating who the perceived hero was, such as "Hero: FL Governor." A codebook of the narrative elements that I was coding for, which include *characters, devil-angel shift,* and *scientific certainty* and *uncertainty* can be found below.

Coding was done by a primary coder; however, to increase reliability, a second coder analyzed every five statements coded. Across codes, intercoder

Table A.6.4 Narrative elements used for data coding.

Code	*Definition*
Characters	
Perceived hero	Those who take action to fix the problem at hand
Perceived villain	Those who create harm upon a victim or one who opposes the aims of the hero
Perceived victim	Those who are harmed by a particular action or inaction
Devil-Angel Shift	
Devil-shift	The message source inflates the "evil-ness" of an entity
Angel-shift	The message source inflates the "good-ness" of an entity
Science	
Use of science	The message source uses science to define a problem, counter a problem definition or justify a policy approach
Scientific uncertainty	The message source questions or disputes the science used by the opposition
Scientific certainty	The message source uses science to establish certainty in their argument

agreement was on average 91 percent.[3] For further quality assurance, the primary returned and double-checked all codes that were assigned three months later.

We performed quantitative analyses and created contingency tables. One set of tables showed the percentages of messages containing each of these codes, and the second showed these percentages broken down by opponents and supporters, with two-tailed hypothesis tests of different proportions. We also utilized the subcodes to provide further analysis of the casting of characters by pro and anti-project stakeholders.

3 80–90 percent intercoder agreement is the benchmark (Saldaña, 2016).

APPENDIX CHAPTER 7

Methods

Case selection and research design

The proposed New Cruise Berthing Facility project in the Cayman Islands was selected as the case study for this research. This port infrastructure project was selected because, in the years following its announcement in 2013, an environmental dispute between two coalitions (over the potential economic benefits versus environmental impact) arose and continues to this day. Both coalitions ran sophisticated campaigns that pushed their policy beliefs to persuade the public and government to proceed with or cancel the project. The project's threat to adjacent coral reefs and an unprecedented outcome of an indefinite halt of the project,[1] mobilized by the non-dominant coalition, make this case of special interest.

This research uses a critical case-study research design to gain in-depth and contextual insights into a complex and contemporary issue (Yin, 2018). This case qualifies as a critical case because understanding how coalitions are formed to contest infrastructure projects that threaten coral reefs is important for future environmental disputes and is not limited to coral reef conservation. A qualitative approach to content analysis is used, wherein the researchers collect document data from local newspapers to identify coalition properties and the occurrences of policy change and learning. The Advocacy Coalition Framework is used with deductive logic to thematically code and analyze data (Saldaña, 2016).

---

1 Coral reefs are globally recognized as vulnerable ecosystems and have strict environmental policies and laws that guarantee their protection from adverse impacts, such as large infrastructure projects, in the Cayman Islands (see Appendix A).

Table A.7.1 International and domestic policies and corresponding governing agencies that protect coral reefs in the Cayman Islands.

Policy/ Plan/ Law	*Agency in Charge*	*Purpose*
International		
Convention on Biological Diversity (1992)	Conference of the Parties[1]	Legally binding international treaty that tasks signatories to develop and implement conservation management plans to preserve species and ecologically important habitat.
Aichi Biodiversity Targets (2011–2020)	Conference of the Parties	Reduce direct pressure on biodiversity, address underlying causes of biodiversity loss across governments, preserve ecosystems and enhance ecosystem services and biodiversity benefits for all.
Cayman Islands Environment Charter (2001)	Cayman Islands Government and United Kingdom Government	This Charter ensures that the environmental resources are extracted sustainably and used wisely, that the risks and benefits are assessed regarding development projects, that environmental impact assessments are performed prior to development projects, the government commits to open consultancy and transparency about development projects, that legislation enforces a "polluter-pays" principle, and that natural heritage is passed down generations through education and preservation.
Domestic		
Marine Conservation Law (1978)	Marine Conservation Board[2]	The purpose is to preserve the marine environment of the Cayman Islands for future generations. There are different zones, including the Marine Park Zone, which prohibits the taking of any marine life. Violations of these laws result in hefty fines or imprisonment or both.
Cayman Constitution (2009)	Cayman Islands Government	This is the highest order of law in the Cayman Islands. Part I Paragraph 18 outlines that the government should protect the environment for future generations while promoting justifical economic development. It should ensure that development is sustainable and limits ecological degradation and promotes conservation.

(Continued)

Table A.7.1 (Continued)

Policy/ Plan/ Law	*Agency in Charge*	*Purpose*
National Biodiversity Action Plan (2009)	Cayman Islands Department of Environment	The goal of this action plan is to have zero extinction in the Cayman Islands while ensuring the protection of natural resources.
Habitat Action Plan (2009)	Cayman Islands Department of Environment	Attain legislative protection status for all coral reefs, expand marine parks and oppose developments that result in coral reef loss.
National Conservation Law (2013)	National Conservation Council[3]	To protect and preserve endemic, endangered and threatened flora and fauna and their respective habitats and prohibits any development that adversely affects a protected area or priority species such as coral reefs.

[1] The Conference of the Parties is the decision-making body for the Convention. All Parties that signed the Convention on Biological Diversity are represented. Together they review and promote the implementation of action plans and convene annually.

[2] The Marine Conservation Board consists of nine members appointed by the Governor, three of which need to be residents of Cayman Brac or Little Cayman (Marine Conservation Law, 1978).

[3] The National Conservation Council has 13 voting members, who come from the Department of the Environment, Department of Agriculture, Department of Planning, and the National Trust. Some members are also appointed by the cabinet based on the geographic districts in the Cayman Islands. Decisions are made via majority vote (Dunning, 2021).

Data collection

A total of 420 news articles were collected as the data for this book. Articles were collected from two local newspapers: The Cayman News Service and the Cayman Compass. Newspaper sources were chosen because the port project overtook national headlines for seven years with stories running almost daily for weeks at a time between 2018 and 2019. The Cayman News Service and The Cayman Compass were chosen because they are the two prominent online local newspapers in the Cayman Islands. While many articles were neutral, the Cayman News Service emphasized the views of and expressed support for the anti-port coalition, while the Cayman Compass (to a lesser extent) supported the pro-port coalition. The earliest article was dated on June 8, 2015 and the latest on September 9, 2021. n = 308 articles exist under the tag "cruise dock facility" in the Cayman News Service, of which n = 283 were included. Those that were excluded often repeated a prior article and did not give new information. n = 201 articles are found under the "Port Vote 2019" issue in the Cayman Compass, of which n = 137 were included. The Cayman News Service articles were collected and coded first, as such there

Table A.7.2 Data collection summary.

	Tag/Keyword	*Number of Articles*	*Date Range*
The Cayman News Services	"Cruise dock facility"	283	June 2015–September 2021
The Cayman Compass	"Port Vote 2019"	137	February 2016–April 2021
Total	NA	420	June 2015–September 2021

were more articles excluded from the Cayman Compass because information was repeated. The entire article was the unit of analysis.

To supplement and triangulate my document data points, we conducted $n = 6$ semi-structured interviews for the Cayman Islands case study, see Table 4.3 for more details (Nightingale, 2020; Price, 2019). To find potential respondents, we used a snowball sampling technique where we asked interviewees to recommend other stakeholders for us to interview (Sadler et al., 2010; Taherdoost, 2018). Interviews were conducted between May 2021 and September 2021. At the beginning of each interview, respondents were asked for permission to record the interview via Zoom. If permission was given, the interview was recorded and transcribed from the recording at a later time. If permission was not given, their responses were transcribed during the interview and edited at a later time, only to improve coherency of the sentences. Respondents were asked nine questions about the costs and benefits of the project, its impact on coral reefs and the use of science regarding the port projects. Interviews ranged from 19 to 127 minutes (interview questions are below). Due to the anonymity and given verbal consent by respondents in these interviews, ethical status was maintained.[2]

Interview Manual

1. What was your role during the cruise berthing facility project dispute?
2. What were/are the benefits of the dredging project?
3. Who is or will be benefiting from them?
4. What were/are the costs of the dredging project?
5. Who is or will be paying for them?
6. In Florida, there are established rules for protecting coral reefs. Did the port dredging project challenge this in any way?

---

2 Protocol # 21-103 EX 2102.

Table A.7.3 Advocacy Coalition Framework Codebook used by researchers.

Code	*Definition*	*Inclusion*
Policy change		
External events	External events refer to any events outside the policy subsystem, which is the port project in this case, that have disrupted its course (i.e., COVID-19).	When the article mentions an outside event that has disrupted the project
Internal events	Internal events refer to any disturbing instances inside the policy subsystem, which is the port project in this case, that have disrupted its course.	When the article mentions an internal event, likely a failure in government, that impacts the coalitions and/or the project
Policy-oriented Learning	Policy-oriented learning can be defined as coalitions and policy actors revising or shifting prior belief systems and intentions resulting from gained experience, exposure, knowledge or for strategic gain.	When the article mentions a revision of beliefs or intentions, mentions any forums to discuss the project

7. How did your organization ensure that science was used to inform the project?
8. How did your organization use social or regular media to portray the project?
9. Did communicating about this project involve significant legal or financial efforts from your organization?

Data coding and analysis

Articles were coded using a codebook based on the Advocacy Coalition Framework. The structural coding methodology was used, in which the researcher codes the data according to the research topics (Saldana, 2016). Structural coding enables the researcher to take a larger data-set, such as a newspaper article, and break it down into smaller components for further analysis. In this case, the components that were coded for included policy change via *external events, internal events, policy-oriented learning.* Data analysis in the form of deductive qualitative content analysis was used as outlined by Elo and Kygnas (2008). Coding each article allowed for qualitative assessment of each component, providing the researcher with the data to understand how the environmental dispute between coalitions unfolded in the Cayman Islands.

APPENDIX CHAPTER 8

Theoretical Framework

To best conceptualize the importance of cultural ecosystem services, and the values they hold for the people who rely on these ecosystems, we must clearly define some of the key terms outlined in our theoretical framework, derived from Fish et al., 2016. In this framework the fundamental building block for understanding how cultural ecosystem services and the benefits they provide work in tandem. It begins with the biophysical domain in which humans interact with their environment. The biophysical domain is defined as nature's condition and ecological functioning (independent of interactions with people/benefits/services). Stated simply, the biophysical domain provides the material components of the environmental spaces where these interactions take place. Additionally, the biophysical domain provides opportunities for the various cultural practices enabled by environmental spaces. The environmental spaces and cultural practices (described in further detail below) both work to shape the biophysical domain. In the Cayman Islands, the reefs in Georgetown Harbour being threatened or lost by the port expansion project are the biophysical domain and the environmental spaces being examined in this chapter.

Environmental spaces are defined as the physical locations or sites in which humans and the societies and cultures they populate, interact within the ecosystem. This can include meadows, streams, mountain tops, and in the context of the Cayman Islands, the beaches and reefs surrounding the islands. These spaces provide a context for the linking of the biophysical domain with cultural practices and cultural ecosystem benefits, as these spaces shape the biophysical domain and enable cultural practices and benefits.

In the case of the Cayman Islands, a majority of the environmental spaces which provide these benefits are reefs, beaches, lagoons and other marine environments surrounding the islands. Popular locations around the three islands (Grand Cayman, Little Cayman and Cayman Brac) include Eden

Rock, Devil's Grotto, Macabuca, Cheeseburger Reef, Seven Mile Beach and the USS *Kittiwake*.[1]

Methods

This study was distributed via an online platform (Google Forms) through the Cayman Islands National Trust which posted the link to the survey on their social media on November 1, 2019. The survey had $n = 175$ responses, which is a very small portion of the Caymanian population, but with information saturation on topics such as the form of reef access that Caymanian's prefer. Criticisms of this research could include a bias since a governmental agency is posting the survey, perhaps only conservation-inclined respondents would provide data. While this may be true, the purpose of the research is not to acquire a representative sample of quantified preferences for policy trade-offs on Caymanian reefs (e.g., what is preferable more conservation or more cruise tourism development) but rather to study how people culturally value Caymanian reefs. All responses were anonymous, with data de-identified. Completion of the survey was voluntary with no incentive and no risk to participants.

Qualitative data in the form of short answer questions were coded using thematic coding, whereby large portions of qualitative text would be boiled down into common themes using a pre-existing framework to categorize themes. The process of coding data adopted deductive reasoning whereby we began with a framework in mind, and assigned codes based on the framework. Four different coders assigned codes to this data, checking every 10th data point to ensure intercoder reliability.

1 Eden Rock, Devil's Grotto, Macabuca and Cheeseburger Reef are dive sites in the Cayman Islands. Seven Mile Beach encompasses several popular dive sites. The USS *Kittiwake* is a shipwreck dive site.

APPENDIX CHAPTER 9

Research Methods

Research design

This case study uses an exploratory sequential research design, where qualitative data is collected first, followed by quantitative analysis to further our understanding of qualitative results (Yin, 2009). The purpose behind exploratory sequential-design studies is that the qualitative findings can inform the quantitative model and in our case become a new theoretical framework for why lawmakers support coral conservation policy (Creswell & Plano Clark, 2018). We used qualitative methods to characterize the messages and themes that political entrepreneurs used to describe their rationale for supporting bipartisan coral reef conservation bills. Quantitative analysis of these statements examines (1) whether there were significant differences between Republicans and Democrats over the ways they framed the problem; and (2) whether support for these bills (i.e., co-sponsorship and "yea" roll-call votes) could be predicted by representatives' jurisdictions being located in coastal zones and/or adjacent to coral reefs. Roll-call votes occur when a representative or senator votes "yea" or "nay," so that the names of members voting on each side are recorded, compared to voice votes which do not record the stance taken by individual members (U.S. Senate, 2021). Thus, the qualitative findings (how lawmakers defined the problem) informs the quantitative analysis (differences in message content across political parties and spatial/political party variables of interest). We added the spatial variables after recognizing that a compelling alternative explanation for support for coral reef legislation may come from the fact that congressional political entrepreneurs may live next to and therefore care more about coral reef ecosystems, regardless of political party or long-term political ideology. Overall, we are theorizing why lawmakers cross party lines to support conservation legislation for climate-vulnerable ecosystems. Understanding this rationale may help enact future bipartisan climate policy.

Qualitative sampling logic

To understand how policy entrepreneurs and political entrepreneurs discussed their rationale for supporting these bills (a process that we conceptualize as "defining the problem in the problem stream"), we collected all statements made on The Restoring Resilient Reefs Act, the Tropical Forest and Coral Reef Conservation Reauthorization Act, and the Offshore Wind for Territories Act. Statements were collected from the Nexis Uni Database, Congressional Record archive on congress.gov, and on the social media platform Twitter for statements made by relevant legislators, NGOs and other members of civil society. We decided to include Tweets because it has been found that politicians use Twitter to advocate their political positions (Hemphill et al., 2013).

The final count of statements made by political entrepreneurs for the Restoring Resilient Reefs Act was $n = 40$. When policy entrepreneurs were included, such as civil society, scientists, agency employees, private industry representatives and NGO representatives, the total number of statements to $n = 70$. We began this Nexis Uni search on February 24, 2020 by searching "restoring resilient reefs act" that yielded $n = 53$ results. Ten of these results contained multiple statements from congressional members, totaling 26 individual statements on the Restoring Resilient Reefs Act from 2019 to 2021. We checked the Congressional Record on congress.gov but there were not any statements made about the Restoring Resilient Reefs Act in the results. Coding began with Tweets made by legislators who co-sponsored the bill, found with an advanced Twitter search for "restoring resilient reefs act" from all the Twitter accounts of the primary sponsors and original cosponsors.[1] This resulted in 13 Tweets to code using the same codes developed from the Nexis Uni database results and iteratively including any new codes into my codebook. A legislative committee hearing on the Restoring Resilient Reefs Act of 2021 was incorporated into the same dataset.[2]

The total number of statements for the Tropical Forest and Coral Reef Conservation Reauthorization Act for policy and political entrepreneurs was

1 Darren Soto, Charlie Crist, Tulsi Gabbard, Jenniffer González-Colón, Brian Mast, Stephanie Murphy, Aumua Amata Radewagen, Marco Rubio, Mazie Hirono, Brian Schatz and Rick Scott.

2 This hearing was held in the House Natural Resources Subcommittee on Water, Oceans and Wildlife Hearing on May 5, 2021. Statements were found on Nexis Uni for both congressional members and expert testimony.

n = 32. Using similar methods[3] to code the Tropical Forest and Conservation Reauthorization Act, 17 documents containing *n = 41* statements were collected from Nexis Uni for this bipartisan legislation. Of these results, 31 were from political entrepreneurs with the remaining 10 being from policy entrepreneurs. Of these, nine documents contained direct statements about the policy, producing 28 individual statements on the Tropical Forest and Coral Reef Conservation Reauthorization Act from 2009 to 2021. We found one additional statement in the Congressional Record and three Tweets to include from the original co-sponsors of all past bills.[4]

The Offshore Wind for Territories had a total of *n = 35* results, with *n = 33* statements from political entrepreneurs and *n = 2* from a policy entrepreneur in the private sector. The same method[5] was used for coding the Offshore Wind for Territories Act statements from relevant policy entrepreneurs. We searched Twitter to find Tweets from all of the past co-sponsors of this bill.[6] We searched congress.gov[7] and added one document of statements from the Congressional Record for this policy.

The combination of all of these messages from policy entrepreneurs brought the total number of messages to *n = 235*. The total number of messages from political entrepreneurs in Congress was *n = 697*. Both sets are mutually exclusive and analyzed separately to understand the difference in how policy and political entrepreneurs framed their messages.

Qualitative coding

For the initial round of qualitative coding, we employed what is known as *in vivo grounded theory coding*, a type of coding that draws on the speaker's own language to develop codes (Saldaña, 2016). This technique was important for understanding how respondents were defining the problem in the problem

---

3 On April 26, 2021, I searched for ["tropical forest and coral reef conservation act"] on Nexis Uni.

4 Representative Steve Chabot, Rep. Brad Sherman, Rep. Jeff Fortenberry, Rep. Eliot Engel, Rep. Christopher H. Smith, Rep. Raul M. Grijalva, Rep. Betty McCollum, Senator Rob Portman, Senator Richard Burr, Senator Sheldon Whitehouse, Senator Tom Udall and Senator Brian Schatz.

5 On May 10, 2021, I searched for "offshore wind for territories act" on Nexis Uni.

6 Representative Darren Soto, Resident Commissioner Jenniffer González-Colón, Delegate Aumua Amata Coleman Radewagen, Delegate Gregorio Kilili Camacho Sablan, Delegate Stacey E. Plaskett, Senator Brian Schatz and Senator Bill Cassidy.

7 We used the terms "offshore wind for territories act," refined by searching "'offshore wind' for territories act," and finally searched for the H.R. bill identifier number in the Congress was introduced to find relevant results in the Congressional Record.

stream. Our research's main interest lies in when and how policy entrepreneurs use the same language or focus on the same characteristics of the problem and propose how to address it through policy-making. Thus, in vivo grounded theory coding assures that the researcher has grasped these most important meanings (Charmaz, 2014). We completed three rounds of coding to (1) initially understand the perspectives of different speakers, (2) categorize those initial codes into discrete categories of the different ways policy entrepreneurs were advocating for these policies, and (3) break these categories down further into specific themes being used to frame these bills on either sides of the aisle. Each individual statement from each speaker received its own entry in the data sheet for each bill. We define an individual statement as those being offset by an indentation, and we chose to code those separately if the individual statements contained opinions with different meanings/emphases or were made at different points in time.

The initial round of coding took place in a word document via bolding important parts of the text (example in Table 1). Codes and their meaning were inventoried in an excel codebook after a process of iterative reviewing of the in vivo codes and an assembly of the initial language into a code of what the root of the statement's meaning was for all respondents (example in Table 2). The codebook included the name of the code, a one to three sentence description of the coded datum's qualities or properties, inclusion criteria, an example of data that best represented the code and possible categories for that code (Bernard & Ryan, 2010). All statements were coded across all three policies.

Following Cresswell's (2013) methods for "lean coding," the 23 initial codes were condensed into eight categories and five final themes. In instances of "fuzzy sets" where categories are discretely separate and some codes overlapped, initial codes were included in the counts and statistics for the categories and themes where overlap occurred (Saldaña, 2016). An example is the initial code for the "economic reality in developed countries" described as an issue of human security and economics in the transcript from policy entrepreneurs. Information saturation was reached for each policy, where we discovered no new information during analysis (Saunders et al., 2018).

Descriptive statistics were employed to analyze these qualitative findings. Frequency tables were constructed to observe the frequency/percentage each theme received for each case policy. One table includes all speakers in civil society (policy entrepreneurs) and the other includes messaging solely from legislators (political entrepreneurs). We examined the aggregate use of the major themes first, ascertaining patterns in use (e.g., most frequently used terms) by calculating population proportion divided by the total number of statements. Then to measure differences in thematic usage between parties, we used an independent, two-tailed, unpaired t-test to determine whether there were significant differences. We used an unpaired independent test since party membership variables are independent of each other and a two-tailed test to look at the differences between the two groups' messaging because no directionality was assumed.

Quantitative logistic model

The overall goal of our research is to theorize rationale for decision-maker support for climate policy, using policy for conservation of climate-vulnerable ecosystems as a multiple case example. To account for alternative explanations for rationale behind lawmaker support, a quantitative model was necessary to analyze competing explanations for support of coral reef policies. We compiled a dataset focused on whether a legislator supported any bipartisan coral reef legislation by co-sponsorship or voting in the House of Representatives for all history. We chose to focus on the House of Representatives for our logistic regression analysis because of their higher dependence on messaging due to their shorter election cycles than the Senate (Lee, 2016).

The first bipartisan bill meeting the criteria for inclusion was introduced in the 101st Congress and the last in the 117th Congress; thus the time frame for the model included 1988–2021. Using an Excel spreadsheet of each member of the House from 1988 to 2021, we coded whether they represent a *coastal state* or a state with coral reefs adjacent to its shoreline (*coral reef state*) as binary variables (Table 3). We created a second variable covering which *political party* the lawmaker belonged to, either Democratic[8] or Republican. We also included descriptive information for the year, session of Congress, state and representative district; however, these were not included in the model.

Variables for the *ideology* of congressional representatives were included based on NOMINATE scores. NOMINATE scores are a measure of ideology based on roll-call votes in Congress with lower scores denoting more liberal ideologies and higher scores denoting more conservative ideologies, thus assuming members' preferences are fixed throughout the course of their career (Lewis et al., 2019). The first dimension (*Ideology 1*) captures economic issues and the second dimension (*Ideology 2*) captures social issues.

Following Harbridge's (2009) rationale for not relying solely on roll-call votes to measure bipartisanship, we also measure *co-sponsorship*, a concept that

8 For the purposes of this research, Rep. Bernie Sanders (VT-I) was coded as a Democrat given his history of caucusing with the party, his seeking the party's nomination for the presidency, and current leadership position within the party.

allows us to see whether a lawmaker attached his or her name to a piece of legislation. This is also important to include because some of these coral reef policies never move forward through committees or pass without a recorded roll-call vote. Without a vote, co-sponsorship is the only opportunity members have to officially document their position on the issue. Co-sponsorship data has previously been used by other researchers to measure ideological positions on bills (Goodliffe et al., 2005; Koger, 2003; Krehbiel, 1995).

Another potential reason why a previous opponent to climate policy is supporting conservation bills for a climate-vulnerable ecosystem is because they're politically vulnerable. To tease these patterns out, we included a variable called *competitiveness*, which measures a lawmaker's vulnerability by looking at the Democratic share of the two-party vote in each of the last three presidential elections. We chose to use the presidential vote instead of the individual member's votes to measure electoral competitiveness of the district. The member's vote share is more reflective of their popularity in the district than the underlying partisan distribution (Fair, 2009). Since House elections occur twice as often as presidential elections, we entered the value for vote share for the most recent past election (i.e., 2016 presidential vote share for members elected in 2018).

To build the dataset for our model, we used congress.gov and used the search term "coral reef" and went through each bill; we chose to exclude resolutions, concurrent resolutions, joint resolutions and amendments since there is less political significance to supporting these types of legislation. After this filter, a total of 183 results were presented, 55 fit the five-part criteria we developed for the model. Bills were included if they (1) enabled the federal government to do something on the coral reef resource itself for conservation, (2) further a federal program, (3) further ideas or science for reef conservation as a main component, (4) address stressors that directly affect reef ecosystems, and (5) had at least one co-sponsor each from each political party. An example of this is H.R. 4900 Oceans Conservation, Education, and National Strategy for the 21st Century Act that would maintain programs for the restoration of marine and freshwater resources through ecosystem-based management. All data included all original sponsors, co-sponsors, and roll-call votes for final passage on bills related to reefs from 1988 to 2021.[9]

9 Data collection ended July 1, 2021 and does not include any subsequent bills that would fit the criteria that were introduced after this date.

INDEX

Ingram Content Group UK Ltd.
Milton Keynes UK
UKHW012207220623
423877UK00003B/10

9 781839 986710